Renewable Energy

SOURCES AND METHODS

GREEN TECHNOLOGY

Renewable Energy

SOURCES AND METHODS

Anne Maczulak, Ph.D.

An imprint of Infobase Publishing

RENEWABLE ENERGY: Sources and Methods

Facts On File, Inc.
An imprint of Infobase Publishing
132 West 31st Street
New York NY 10001

Library of Congress Cataloging-in-Publication Data
Maczulak, Anne E. (Anne Elizabeth), 1954–
    Renewable energy : sources and methods / Anne Maczulak.
        p. cm.—(Green technology)
    Includes bibliographical references and index.
    ISBN 978-0-8160-7203-3
    1. Renewable energy sources—Juvenile literature. I. Title.
    TJ808.2.M33 2010
    621.042—dc22   2009010352

Text design by James Scotto-Lavino
Illustrations by Bobbi McCutcheon
Photo research by Elizabeth H. Oakes
Composition by Hermitage Publishing Services
Cover printed by Bang Printing, Brainerd, MN
Book printed and bound by Bang Printing, Brainerd, MN
Date printed: October 2010
Printed in the United States of America

10 9 8 7 6 5 4 3 2

Contents

Preface   ix

Acknowledgments   xi

Introduction   xiii

1 Earth's Energy Sources   **1**

The World's Appetite for Energy   2

Case Study: Western Energy Crisis, 2000–2001   6

Renewable or Nonrenewable   8

Oil   10

Sun's Energy Stored in the Earth   14

Carbon Economics   16

Our Renewable Energy Resources   20

Smart Energy Grids   22

Social Aspects of Alternative Energy   24

Energy Programs in the Global Community   26

The U.S. House Subcommittee on Energy and Environment   28

Conclusion   28

2 Recycling   **31**

The Grassroots History of Recycling   33

How Recycling Saves Energy   35

Industrial Raw Materials from Waste   39

Recycled Materials Chemistry   41

High-Density Polyethylene (HDPE)   42

Minerals and Metals   44

Case Study: Recycling during World War II 46
Rubber Recycling 48
Conclusion 49

3 Gasoline Alternative Vehicles **51**

Evolution of Alternative Vehicles 53
Biofuels 57
Case Study: Toyota's Prius 58
Synthetic Fuels 62
Battery Power 64
Combustion 65
Fuel Cell Technology 66
Nuclear Fission and Fusion 69
Natural Gas Fuels 70
Next Generation Hybrids 72
Conclusion 74

4 Biorefineries **75**

Today's Refinery Industry 77
Pipelines 80
The U.S. Department of Energy 81
Case Study: Alaska's Oil Economy 84
Biorefining Steps 86
Developing the Biorefining Industry 89
Conclusion 91

5 Innovations in Clean Energy **93**

Alternative Energy Emerging 94
Wind, Wave, and Tidal Power 97
Solar Power 101
Solar Films 108
Hydropower and Geothermal Energy 109
Nuclear Energy 113
Direct Carbon Conversion 117

Fuel Cells 120
Conclusion 124

6 Green Building Design **125**

Building Green Comes of Age 127
Controlling Energy and Heat Flows 129
Leadership in Energy and Environmental Design
   (LEED) 131
Cooling and Ventilation 138
Insulation 140
Daylighting 142
Windows Technology 144
Water Conservation 146
Case Study: Four Horizons House, Australia 147
Managing Waste Streams 150
Off the Energy Grid 151
Conclusion 154

7 Energy from Solid Biomass **156**

The Earth's Biomass 158
Types of Biomass 160
The Phosphate Bond 163
Conversion to Energy and Fuels 164
 The Energy Value of Garbage 166
A Biomass Economy 167
Case Study: The Chicago Climate Exchange 169
Conclusion 171

8 Future Needs 172

Appendixes 174
Glossary 184
Further Resources 190
Index 199

Preface

The first Earth Day took place on April 22, 1970, and occurred mainly because a handful of farsighted people understood the damage being inflicted daily on the environment. They understood also that natural resources do not last forever. An increasing rate of environmental disasters, hazardous waste spills, and wholesale destruction of forests, clean water, and other resources convinced Earth Day's founders that saving the environment would require a determined effort from scientists and nonscientists alike. Environmental science thus traces its birth to the early 1970s.

Environmental scientists at first had a hard time convincing the world of oncoming calamity. Small daily changes to the environment are more difficult to see than single explosive events. As it happened the environment was being assaulted by both small damages and huge disasters. The public and its leaders could not ignore festering waste dumps, illnesses caused by pollution, or stretches of land no longer able to sustain life. Environmental laws began to take shape in the decade following the first Earth Day. With them, environmental science grew from a curiosity to a specialty taught in hundreds of universities.

The condition of the environment is constantly changing, but almost all scientists now agree it is not changing for the good. They agree on one other thing as well: Human activities are the major reason for the incredible harm dealt to the environment in the last 100 years. Some of these changes cannot be reversed. Environmental scientists therefore split their energies in addressing three aspects of ecology: cleaning up the damage already done to the earth, changing current uses of natural resources, and developing new technologies to conserve Earth's remaining natural resources. These objectives are part of the green movement. When new technologies are invented to fulfill the objectives, they can collectively be called green technology. Green Technology is a multivolume set that explores new methods for repairing and restoring the environment. The set covers a broad range of subjects as indicated by the following titles of each book:

- *Cleaning Up the Environment*
- *Waste Treatment*
- *Biodiversity*
- *Conservation*
- *Pollution*
- *Sustainability*
- *Environmental Engineering*
- *Renewable Energy*

Each volume gives brief historical background on the subject and current technologies. New technologies in environmental science are the focus of the remainder of each volume. Some green technologies are more theoretical than real, and their use is far in the future. Other green technologies have moved into the mainstream of life in this country. Recycling, alternative energies, energy-efficient buildings, and biotechnology are examples of green technologies in use today.

This set of books does not ignore the importance of local efforts by ordinary citizens to preserve the environment. It explains also the role played by large international organizations in getting different countries and cultures to find common ground for using natural resources. Green Technology is therefore part science and part social study. As a biologist, I am encouraged by the innovative science that is directed toward rescuing the environment from further damage. One goal of this set is to explain the scientific opportunities available for students in environmental studies. I am also encouraged by the dedication of environmental organizations, but I recognize the challenges that must still be overcome to halt further destruction of the environment. Readers of this book will also identify many challenges of technology and within society for preserving Earth. Perhaps this book will give students inspiration to put their unique talents toward cleaning up the environment.

Acknowledgments

I would like to thank the people who made this book possible. Appreciation goes to Bobbi McCutcheon who helped turn my unrefined and theoretical ideas into clear, straightforward illustrations. Thanks also go to Elizabeth Oakes for providing photographs that recount the past and the present of environmental technology. My thanks also go to Jodie Rhodes, who is a constant source of encouragement. Finally, I thank Frank Darmstadt, executive editor, and the talented editorial staff at Facts On File for their valuable help.

Introduction

Hundreds of generations have relied on a fairly short list of energy sources to perform work. Wood, coal, oil, and natural gas served well as fuels for providing heat and for cooking; wind and water powered sawmills and ships at sea. But the human population did not confine itself to a simple life. Communities expanded and needed new types of vehicles. These communities soon began growing at a pace that outstripped natural resources. Some parts of the world ran out of resources faster than other regions, but they maintained their strong, growing economies by importing materials from resource-rich areas. Forests began disappearing, and challenges in mining coal increased. Crude oil reserves also became increasingly difficult to find and tap, and eventually scientists could calculate a future point in time when the world's oil reserves would be gone.

People noticed something else as early as 1950: The skies had turned heavy with pollution. Technology had certainly brought new conveniences to civilization, but these technologies also introduced problems to a generation left with the job of finding ways to halt the harm being done to the environment. Graduates in physics, chemistry, engineering, biology, and ecology would soon be asked to accept the job of reinventing the way society used and reused materials for making energy.

Renewable energy offers an advantage compared with other disciplines in environmental science because of the breadth of new technologies emerging every day in this field. Although U.S. energy technologies once centered squarely on extracting coal and crude oil, new technologies began to contribute to overall energy consumption. The *nuclear energy* industry grew in the 1950s but over time its promise dimmed. Nuclear power's future remains very uncertain, and, as communities have resisted nuclear energy for a variety of safety concerns, coal, oil, and natural gas again dominate world energy production—these three energy sources supply 87 percent of energy needs worldwide. Since the 1970s, world energy production has been led by mammoth oil companies, coal producers, and power utilities that use either nuclear or nonnuclear energy sources.

The first warning of a sea change in world energy supply occurred in the 1970s when a newly established oil *cartel* in the Middle East took control of the region's plentiful supply of cheap oil. Americans learned to adjust to new speed limits and higher prices. Later, drivers contended with gas rationing in which fuel purchases were limited. Perhaps the gasoline pump would someday run dry.

As the U.S. oil supply from foreign shores began to look a bit less certain, environmental experts interjected more disquieting news. They warned, first, that the Earth's oil supply would reach a point of diminishing returns and, second, that fuel emissions were accumulating to dangerous levels in the atmosphere, enough to cause global temperatures to rise. The public found it difficult to imagine that an extra car trip to the store could in some way make the Earth's temperature rise. Many people therefore ignored the impending global climate crisis and continued driving far and fast, encouraged by the fact that the scientific community was immersed in a heated debate over whether global warming truly existed.

In the 1990s, Vice President Al Gore spoke for a growing consortium of scientists who had been collecting evidence of rising temperatures in the atmosphere. They warned the public that emissions from all forms of transportation, certainly a large portion from cars and trucks, were accumulating in the atmosphere and interfering with the Earth's normal heating and cooling cycles. By the end of the decade, a small number of automakers offered drivers a new opportunity, that is, a car powered by a dual gasoline-electricity system. The number of experts concerned over Earth's warming trend increased, and the overwhelming majority of them warned that climate change was caused not by nature but by humans. Some drivers tried the new gasoline-electric vehicles and found that they conserved gasoline and cut emissions. But this shift in thinking did not sway most U.S. car buyers or big U.S. automakers that equated driving with heavy vehicles with big engines, driven at high speeds.

It is difficult to identify a defining moment that turned the tide toward less-polluting energy sources, but by the start of the new century most people had developed a fresh outlook on the environment. The numbers of people who agreed that the planet might indeed be warming passed the numbers of skeptics. A new community emerged: People who wanted alternative energies for their cars, public vehicles, and houses. Renewable energy sources that bypassed the need for *fossil fuels* became more than a curious idea for staunch environmentalists; renewable energy joined the mainstream. Politicians who once scoffed at the notion of a warming planet changed course and figuratively wrung their hands over the problem of global warming. Today, any politician would be foolish to run for office without first devising a clear and feasible energy plan for conserving fossil fuels.

In 1988, the World Meteorological Organization (WMO) and the United Nations Environment Programme (UNEP) established a team of scientists, governments, and policy experts called the Intergovernmental Panel on Climate Change (IPCC). The IPCC has taken the lead in assessing the current knowledge on global temperatures and *greenhouse gas* buildup in the atmosphere. Ordinary citizens found climate change such a complex issue to grasp that they often overlooked the real evidence before them: rising sea levels, dying forests, an increase in infectious diseases, and debilitated ocean ecosystems, to name a few clues. The IPCC has drawn together all of these issues and communicates the problems, the unknowns, and the possible plans for bringing global warming under control.

Renewable energies, and in particular low-emission energies, make up a crucial part of the IPCC's proposals on climate change. By reading the organization's periodic reports on climate change, a nonscientist quickly learns that no single area of expertise will solve global warming. It is a massive problem caused by a tremendous increase in industrial activities that began with the industrial revolution. But advances in renewable energy hold the greatest potential for affecting a planet that humanity has altered.

Renewable Energy reviews the current status of renewable energy technologies, a critical subject since the world now increases its energy consumption between 1 and 3 percent every year. It covers the current rate of energy consumption and the consequences of continuing at this rate. The book explains how the main conventional forms of energy—coal, oil, and gas—contribute to economies, but its main theme is the remarkable diversity of ideas that are born every day in the field of alternative energy.

The first chapter reviews the Earth's energy sources from the fossil fuels that have been depended upon for the past century to a new approach to energy production and distribution. The second chapter discusses how recycling plays a part in energy conservation by managing natural resources and allowing industries to find new uses for common materials and products. Chapter 2 also covers new technologies in recycling and some of the areas where recycling can improve.

Chapter 3 covers the important topic of alternative fuel vehicles, which will certainly be a crucial piece of a new energy future. The chapter explains why new vehicles cannot be designed and produced as an isolated task, but rather their success depends on cooperation between automakers, fuel companies, and the public. The chapter also describes the technologies behind biologically based fuels, synthetic fuels, batteries, and fuel cells, one of the newest technologies in alterative energy.

Chapter 4 provides information on the biorefining industry, which makes new fuels for transportation or heating by using plant-made compounds, mainly ethanol. It discusses also the promise of biorefining and the daunting challenges ahead for this industry if it hopes to catch up with the fossil fuel industry. Chapter 3 also provides a special look at pipelines in fuel transport.

Chapter 5 describes innovations in clean energy sources that are being pursued mainly because they do not cause as much air pollution as fossil fuels. The chapter explains the advantages and disadvantages of solar, wind, water, and geothermal energies. It also follows the book's theme in emphasizing the incredible number of options and new ideas emerging in these energy technologies.

The next chapter provides a primer on new methods for constructing buildings that are designed for energy and resource conservation. New fuels for vehicles may be consumed in the future at rates that exceed their production. New buildings confront the same challenge; future construction projects will increasingly adhere to principles that reduce waste, reuse materials whenever possible, and construct a building that will conserve energy. Chapter 6 covers the latest technologies in heating and cooling, lighting, insulation, windows, and waste management.

Chapter 7 describes the use and the predictions for *biomass* as an important energy source to conserve fossil fuels. It covers the nature of biomass, why it acts as an energy source, and the decisions that can be made today to make biomass a valuable energy source for the future. The chapter also discusses the emerging business of buying and selling carbon in international markets, surely one of the most innovative ideas to emerge from the renewable energy arena.

Renewable Energy offers an encouraging array of technologies for both the near future and long-term planning. If only half of these new technologies come into being, society will have created a very good chance at saving the Earth from its dangerous course.

EARTH'S ENERGY SOURCES

nergy is the primary force in the universe. Energy defines the Earth's *biomes* and sustains life. All life, from single-celled *microbes* to blue whales, exists in a continuous process of consuming, using, and storing energy. Human communities work in the same way as other communities with regard to energy management. Any community consumes fuel to produce energy, but the community must also conserve some of the fuel for the next generation. This *conservation* of energy sources from one generation to the next is the principle behind *sustainability,* the process by which a system survives for a period of time. No system in biology lasts forever, and this is also true for sustainability. Sustainability prolongs the time that living things can survive, but it cannot ensure that life will go on forever.

The Earth's resources can be called its *natural capital.* Capital is any asset that has value. Natural capital, meaning things in nature such as trees, rivers, coal, and wildlife, must be managed in the same way that responsible people manage their money. A person who possesses $10,000 but spends every penny of it in a single month has not conserved monetary capital. That person certainly will not be able to sustain a comfortable lifestyle. By keeping a budget and making prudent purchases, the same amount of money will last far longer; this is conservation.

A savings account containing $10,000 with no other form of income represents a *nonrenewable resource.* Once the money has been spent, no more money will magically appear. In terms of natural capital, Earth's main nonrenewable resources are oil, natural gas, coal, metals, minerals, and land. Nonrenewable resources can be thought of as depleted when the energy needed to extract them from the Earth costs more than the energy value of the resource itself.

A person can conserve $10,000 by getting a job and earning money to renew any funds spent each month. In the same way, the Earth contains *renewable resources* that replenish over time: forests, plants, wildlife, water, clean air, fresh soil, and sunlight. Renewable resources may take a long time to replenish—forests can take 100 years to mature—or a short time, such as sunlight that returns each morning.

Living sustainably means conserving nonrenewable resources by intelligent use of renewable resources. Even renewable resources must be managed carefully or else they too can disappear faster than they are replaced. The world is now experiencing this very problem because in many places forests, plants, wild animals, clean water, clean air, and rich soil have become depleted before nature can replace them.

Sustainable use of resources depends on the principles of conservation and resource management. Since the 1960s, some people have known that conservation of nonrenewable energy sources is of paramount importance. At the same time, people must put increased effort into using renewable energy sources from the Sun, wind, and water. This chapter examines the renewable energy sources available today, aspects of managing these sources, and new technologies that will be crucial for future generations.

This chapter reviews the current state of energy use in the world and covers specific characteristics of renewable and nonrenewable energies. It covers the ways in which society has come to rely on oil. It contrasts such dependence on fossil fuels with the benefits of switching to renewable energy sources. The chapter also includes special topics related to energy use such as carbon management and the mechanism by which utility companies distribute energy to consumers.

THE WORLD'S APPETITE FOR ENERGY

World energy consumption has increased rapidly since the industrial revolution introduced mechanized production methods. However, since the first Earth Day on April 22, 1970, the public has grown increasingly conscious of the environment and the need to be prudent in the use of natural resources. The rate of energy consumption has slowed in the United States since the 1980s, but Americans continue to use energy lavishly compared with every other part of the world.

Americans consume about 100 quadrillion *British thermal units* (Btu) yearly. (An engine burning 8 billion gallons [30 billion l] of gasoline produces about 1 quadrillion [10^{15}] Btu of energy; 1 Btu equals the energy released in burning one wooden match.) The United States consumes more energy than it produces, so it must import the difference. The following table shows how the United States currently uses its energy sources.

Electric power production uses the biggest proportion, almost 40 percent, of all energy used in the United States. Transportation consumes 27.3 percent of the country's energy use, industry uses 18.8 percent, and residential and commercial buildings use 10.6 percent.

Lifestyle and a country's type of economics affect the rate at which residents consume energy. Many of the highest energy consumers produce very little of the energy they use within their borders. Luxembourg, for example, consumes a rather large amount of energy per capita, yet it produces almost none of its energy. The countries that in the past decade have consistently used the most energy per capita per year are the following:

U.S. ENERGY CONSUMPTION (2010)		
ENERGY SOURCE CONSUMED	PERCENT OF TOTAL ENERGY CONSUMPTION	MAIN SECTORS USING THE ENERGY SOURCE
petroleum	35.3	transportation; industry; residential and commercial; electric power production
natural gas	23.4	transportation; industry; residential and commercial; electric power production
coal	19.7	industry; residential and commercial; electric power production
nuclear power	8.3	electric power production
renewable sources	7.7	transportation; industry; residential and commercial; electric power production
Source: Energy Information Administration (EIA)		

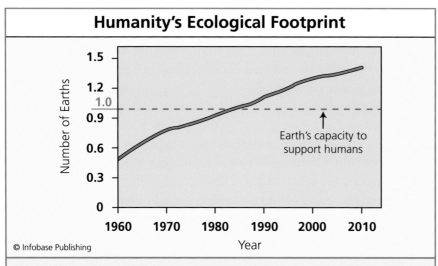

Humanity's Ecological Footprint

Number of Earths

Earth's capacity to support humans

Year

© Infobase Publishing

The world's human population has exceeded its ecological footprint by about 20 percent. Each year, the average person on Earth uses more resources and produces more wastes than the planet can produce or absorb, respectively. Some of the consequences of exceeding the ecological footprint have already become evident: depleted fisheries, diminished forest cover, scarcity of freshwater, and buildup of wastes.

Qatar, United Arab Emirates, Bahrain, Luxembourg, Canada, and the United States. The United States (as a nation) consumes more than 21 percent of all the energy consumed globally; China is the next biggest consumer at 15 percent. Appendices A and B list the top energy-consuming (crude oil) countries and energy consumption trends, respectively.

National energy appetites correlate with a country's industrialization. The International Energy Association (IEA) has estimated that developed countries use about 3.4 million tons (3.1 million metric tons) of energy sources (on an energy-equivalent basis with oil), but developing countries use only 1.7 million tons (1.5 million metric tons).

The world's consumption of renewable and nonrenewable natural resources can be expressed by a calculation called the *ecological footprint*. An amount of energy as large as 100 quadrillion Btu is difficult to imagine, but an ecological footprint puts resource consumption into understandable terms. The ecological footprint equals the amount of land and water needed to sustain life and absorb wastes. This can be calculated for a single person, a country, or the entire planet. Since the mid-1980s, the world's population has exceeded its ecological footprint. In other words, people are consuming resources faster than the Earth can replace them. People are able to notice the effects of a growing ecological footprint when they see polluted water and air, shrinking forests and grasslands, or increasing gas and electricity costs.

The type of energy sources used by society affects the ecological footprint in two different ways. First, some resources require that the land be disturbed to extract the resources, which produces large amounts of dangerous waste. For example, coal mining companies sometimes remove entire mountaintops to get at the coal underneath, and then coal burning puts emissions into the air that cause *global warming*. Second, by reducing the use of resources that damage and pollute the environment and replacing them with renewable and nonpolluting resources, people can reduce their ecological footprints. At this point in history, every individual's goal should be to reduce their ecological footprint as much as possible while maintaining an acceptable lifestyle.

Countries reduce their ecological footprints in the same way as people. Countries can minimize dependency on fossil fuels, encourage the

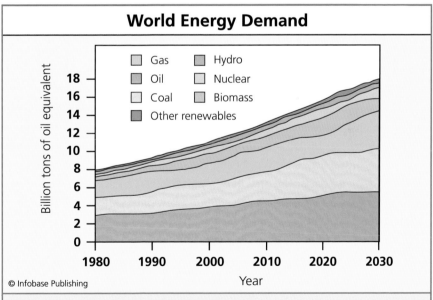

Oil, natural gas, and coal have provided the bulk of world energy consumption since the Industrial Revolution. Of nonfossil fuel energies, nuclear power currently produces about 6 percent of world energy demand, biomass combustion 4 percent, and hydroelectric dams 3 percent. Renewable energies can only have a meaningful effect on ecological footprints if people simultaneously reduce their use of fossil fuels.

development of resources as alternatives to fossil fuels, perfect pollution cleanup methods, and design technology that reuses most of the waste materials that society produces. Countries also must overcome obstacles from pol-

CASE STUDY: WESTERN ENERGY CRISIS, 2000–2001

In the warm summer of 2000, the western United States experienced a sudden increase in energy prices, power outages, and power rationing by utility companies. Many families struggled through *rolling blackouts* in which utility companies rationed electricity to conserve a faltering power supply. In July, a Federal Energy Regulatory Commission (FERC) news release assured that the commission "ordered its staff to conduct an investigation of electric bulk power markets so that it can determine whether markets are working efficiently and, if not, the causes of the problems." Bulk power markets referred to buyers and sellers of electricity throughout the country. FERC's statement began the unveiling of a serious flaw in the U.S. energy supply that would have an impact on financial markets around the world.

The western energy crisis of 2000–2001 began with a drought that lowered water levels and reduced the amount of electricity that *hydroelectric power* plants could produce. Electricity reserves at California's energy utilities fell to low levels, so these companies bought more electricity from Washington and Oregon, which had excess amounts. At the same time, the nation's wholesale supply of electricity called the *power grid* furnished irregular amounts of electricity at varying prices. A crisis began to heighten. California electric companies were required by law to charge no more for electricity than certain predetermined prices. Companies in other states that controlled the grid, however, charged whatever price they liked. California's electricity costs rose higher and higher.

California stumbled into its energy crisis because in 1998 the state deregulated its electric industry, meaning electric supply and distribution to customers occurred on a competitive supply and demand basis. The goal of deregulation was to decrease overall energy costs for customers. But the electricity shortage in 2000 made it difficult for California to buy cheap electricity to keep inexpensive energy flowing. Utility companies began purchasing electricity at high prices as out-of-state electricity sellers knew they could take advantage of California's problem. The prices in short-term electricity sources, called *spot markets*, fluctuated, and energy availability in the state turned into a day-to-day emergency.

California might have weathered its energy shortfall until the autumn, but the out-of-state companies controlling electricity supply increased the pressure even more. Electric wholesalers such as Reliant Energy, Dynegy, and Enron began illegally manipulating electricity prices and supply. Many wholesalers created false data to imply that they too had run short of electricity so

itics, international relationships, and the state of their economies. The sidebar "Case Study: Western Energy Crisis, 2000–2001" on page 6 describes how these factors affect a country's ability to control its ecological footprint.

that they could claim that they were forced to raise prices. Incredibly, these companies devised plans to sell the same electricity over and over and invented schemes that further skewed normal supply and demand patterns. The FERC investigation later explained in summarizing its findings, "One scheme in particular ... is designed to create an illusion of power flowing in a circle from John Day in Oregon to Mead in Nevada [large energy utilities], through the critical congested [electricity] pathways in California, without any input of power whatsoever." Enron and similar companies sold electricity, but they failed to deliver it. By 2001, Pacific Gas and Electric in northern California filed for bankruptcy, and Southern California Edison needed emergency help to avoid the same fate.

Rolling blackouts continued through the winter and into 2001. FERC's investigation meanwhile had made progress in untangling a complicated and multibillion dollar world of buying and selling electricity. By the end of 2001, FERC had collected evidence of numerous irregularities by the energy brokers who had held California hostage to high-energy prices. Further investigations led to the downfall of several wholesale energy companies and jail time and fines for their executives. Various leaders proposed that the federal government take over the nation's energy supply, but the National Energy Development Task Force refused to stop deregulation because the task force felt deregulation made the economy stronger.

The sight of wealthy executives pleading their cases to judges caught the nation's attention more than the root cause of California's energy crisis. The crisis illustrated the importance of maintaining a steady, reliable energy supply to households, but the nation's massive power grid also required diligent control. California had learned several lessons as well. It had failed to build enough power plants to keep up with population growth, and no one had anticipated an unusually cold winter in 2000 or a severe drought that summer that lowered water levels in reservoirs. Hydroelectric plants generated less energy due to the lowered water levels, but demand for power rose, and the crisis also made prices for natural gas rise.

The western energy crisis ruined careers, destroyed companies and jobs, and wiped out billions of dollars in retirement savings. It demonstrated the intricate ways in which energy has become woven into state and national economies. Any future sustainable practices in energy use will need the support and oversight of government and good business decisions.

RENEWABLE OR NONRENEWABLE

The concept of renewable versus nonrenewable resources provides the cornerstone of sustainability. Renewable resources are replaced by natural processes over time, but even these must be conserved so that they are not used up faster than nature can replace them. Conversely, nonrenewable resources such as oil or minerals are formed in the Earth over millions of years. Earth can replenish nonrenewable resources, but this occurs over eons such as the millions of years needed to transform *organic* matter into fossil fuels. Do people have any real chance to affect the entire planet and preserve its natural wealth? Environmentalists think everyone can indeed make a difference in building sustainability by following the three rs— reduce, reuse, and recycle. These activities conserve both renewable and nonrenewable resources, as described in the following table.

Energy companies would be wise not to deplete resources faster than the Earth replaces them, a process known as *recharging*. However, replenishment of renewable resources has become increasingly difficult because of a growing world population. Although many factors contribute to

The world's countries differ in energy use, per country and per capita. This satellite image of the planet's city lights shows where most energy is consumed. In general, countries that produce a large volume of goods and services (high gross domestic product) also consume the largest amounts of energy. The U.S. Department of Energy has joined other energy agencies in projecting that world energy consumption will double in the next 50 years. *(NASA)*

| RENEWABLE AND NONRENEWABLE RESOURCES ||
RENEWABLE RESOURCES	HOW THESE RESOURCES REPLENISH THEMSELVES
air	Earth's *respiration* and plant and animal respiration
animals	reproduction
forests	reproduction and germination
grasses and plants	reproduction and germination
microbes	sexual and asexual reproduction
nutrients (carbon, nitrogen, phosphorus, sulfur, etc.)	decomposition of plant and animal wastes followed by *biogeochemical cycles*
soil	Earth's *sediment cycle*
sunlight	activity at the Sun's core
water	biological reactions, including respiration
wind	climate, tides, and weather
NONRENEWABLE RESOURCES	HOW THESE RESOURCES BECOME DEPLETED
coal	mining for energy production by burning
land	development for population expansion
metals	mining for industrial use
natural gas	extraction for energy production by burning
nonmetal minerals	mining and other extraction methods for industrial use and other commercial uses
petroleum	extraction for energy production by *combustion* and industrial uses
uranium	*nuclear energy* production

population growth at unsustainable rates, two important historical developments may have had the largest impact on population because they increase life span. First, the development of the microscope 275 years ago led to greater knowledge of microbes and an increasing understanding of disease. Second, conveniences introduced by the industrial revolution alleviated the need for manual labor in many industries. In short, life had become less physically demanding, and medicine had reduced the infant mortality rate and lengthened life spans. Populations in developed and developing regions began to undergo *exponential growth,* which means that the numbers of humans increase at an increasingly faster pace over a short period of time.

Exponential population growth is the single most significant factor in humans' increasing ecological footprint. In this decade, humans have been depleting resources 21 percent faster than Earth can recharge them. Environmental scientists often describe this problem as the number of planet Earths that people need to support their activities. At present, humans need 1.21 Earths to support current consumption of resources.

OIL

Crude oil, also called petroleum, is a thick liquid found in underground rock formations. The petroleum industry extracts crude oil out of the ground and then refines it into products such as gasoline. Crude oil contains a complex mixture of compounds made of carbon chains with hydrogen molecules attached to each link in the chain. Extracted crude oil also contains small amounts of sulfur, oxygen, and nitrogen compounds mixed with the *hydrocarbons.* The principle of oil refining is to remove crude oil's impurities, that is, anything that is not a hydrocarbon.

Oil refineries clean up crude oil by heating it to drive off the impurities. This heating step to purify a liquid is called distillation. Light, volatile (easily vaporized) materials such as gases leave crude oil first, and the least volatile components, such as asphalt, remain in the mixture the longest. Refineries recover the following components from crude oil, listed from the most to the least volatile: gases, gasoline, aviation fuel, heating oil, diesel oil, *naptha solvents,* greases, lubricants, waxes, and asphalt.

Refineries further distill some of the components to collect specific chemicals called petrochemicals. Different industries have a need for

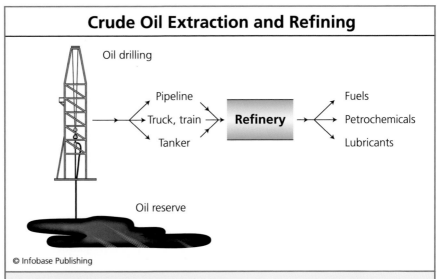

Crude Oil Extraction and Refining

Oil drilling

Pipeline

Truck, train → **Refinery** →

Tanker

Fuels

Petrochemicals

Lubricants

Oil reserve

© Infobase Publishing

Crude oil exploration, extraction, and refining make up a multitrillion dollar industry. The United States has 33 oil refineries employing more than 65,000 people. Service stations employ another 100,000 workers. Any transition to new fuels must be coordinated with the oil industry in order to protect world economies. Many scientists have considered new uses for refineries, perhaps by modifying operations to make natural gas or biomass fuels.

specific petrochemicals that vary mainly by the types of hydrocarbons they contain. Petrochemicals currently go into the production of the following materials: organic solvents, pesticides, plastics, synthetic fibers, paints, and some medicines.

Present global crude oil reserves still contain enough to last several decades. At some point, however, finding new reserves, drilling, and extracting the oil will not occur fast enough to meet the world's oil appetite. The United States reached that critical point about 1970 when crude oil production for the first time stopped increasing in this country and began declining. The United States turned to imports from Saudi Arabia, Mexico, Canada, Venezuela, Nigeria, and Iraq, plus small amounts from other countries, to make up the difference. Overall, U.S. oil supplies come from the places listed in the following table. The rest of U.S. oil requirements comes from domestic sources mainly in the Gulf of Mexico. Of all U.S. oil production, the Gulf of Mexico supplies more than twice the amount of oil than any other region.

SOURCES OF U.S. CRUDE OIL, 2008

Region	Percent of Total U.S. Consumption	Main Individual Contributing Countries
major sources		
North America	36	Canada; Mexico; United States
South America	22	Venezuela; Brazil; Colombia; Ecuador
Middle East	21	Saudi Arabia; Iraq; Kuwait
Africa	17	Nigeria; Angola; Algeria; Congo (Brazzaville)
Europe	4	Russia; Norway
minor sources		
Asia	less than 2	China; Vietnam; Azerbaijan
Oceania (Pacific islands)	less than 1	Australia
Source: Energy Information Administration (EIA)		

The exact amount of crude oil in the world's oil reserves has not been determined even though experts have tried various methods to calculate the remaining volume. Author Sonia Shah pointed out in her 2004 book *Crude: The Story of Oil,* "The size of oil reserves is generally calculated by reservoir engineers employed by oil companies." Once oil companies determine what they believe is an accurate calculation of an oil reserve's size, these companies may be reluctant to make the information public for the three following reasons: (1) to protect the status of their country's oil import or export activities, (2) to better control fuel prices, and (3) to protect national security.

Estimating an oil reserve starts by drilling an appraisal well to gauge the extent of the underground reserve and to sample underground rock for geologists to study. Geologists can make predictions on the probability of finding oil based on the constituents of rock. "Yet even with the most sensitive statistical tests and the most advanced petrochemistry," Shah wrote, "what the oily samples on the lab table reveal about the formations under the ground is limited." Petrochemistry is a specialty in chemistry related to the characteristics of compounds found in crude oil. Shah also quoted petroleum geologist Robert Stoneley of the Royal School of Mines in London, England: "Until we have actually produced all of the oil that we ever shall, we are involved with a greater or less degree of uncertainty." To complicate matters, oil companies change their findings over time as they use more sophisticated methods to calculate oil reserves, and countries may hedge the truth about their oil reserves for political reasons.

Despite the unknowns regarding oil reserves, scientists and nonscientists can agree on the following features of world crude oil:

- The following nations hold the largest oil reserves, in order: Saudi Arabia, Canada, Iran, Iraq, United Arab Emirates, Kuwait, Venezuela, Russia, Libya, and Nigeria.
- The United States consumes the most oil (almost 21 million barrels per day), about three times the next largest consumer.
- China and Japan consume the next largest volumes, more than 7 and 5 million daily barrels, respectively.
- Saudi Arabia holds the largest oil reserves, about 262 billion barrels, followed by Canada with about 180 billion barrels.
- The U.S. oil consumption gap is increasing, which makes the country more dependent on oil imports and fuels that replace oil.

Of all countries in the world, the United States has the largest and fastest-growing consumption gap, calculated as follows:

oil consumption – oil production = consumption gap

China follows closely behind the United States in oil consumption, and since 1993 China has also become an oil importer because its reserves cannot meet its demand. Even the vast oil fields in eastern China have been declining since 1980. As oil-producing countries find their oil reserves more and more difficult to reach, alternative fuels become a critical need. The decision to emphasize alternative and renewable energy sources therefore can be attributed to two factors: (1) the pollution caused by burning petroleum fuels, and (2) the inevitable decline of oil reserves.

SUN'S ENERGY STORED IN THE EARTH

The energy stored in the Earth's crude oil originally came from the Sun. Over thousands of years, generation upon generation of all types of life on Earth thrived, died, and then decomposed. The decomposed organic matter accumulated under the Earth's oceans and migrated into deep sediments. The Earth's mantle exerted tremendous pressure on these organic compounds and the carbon-hydrogen substances became liquid—the oil reserves people depend on today. Humans cannot replicate the process by which Earth formed crude oil, but they can develop other ways to take advantage of the ultimate source of all the energy, in all its forms, on Earth today, the Sun.

Life on Earth uses the Sun's energy either in an indirect or direct manner. The world's oil reserves have stored the Sun's energy for millions of years as a complex mixture of carbon compounds. When people use refined petroleum products to run engines, they are using the Sun's energy indirectly. By contrast, a house heated by sunlight coming in through windows is using the Sun's energy in a direct manner.

Energy is the ability to do work. Walking, typing on a keyboard, and heating a room are examples of actions that require energy. Civilization has devised ways to use the Sun's energy that the Earth stores in the following six forms:

- electrical energy from the flow of electrons
- mechanical energy in things such as engines
- light or radiant energy from the Sun

- heat
- chemical energy in the bonds that hold matter together
- nuclear energy in the nuclei of atoms

Sunlight travels to Earth in the form of energy called electromagnetic radiation. Electromagnetic radiation moves through space at the speed of light, 186,000 miles per second (300,000 km/s), and behaves like a wave in a pond with troughs and peaks. A wavelength is the distance from peak to peak or trough to trough in any type of wave. Sunlight contains a range of wavelengths in which each corresponds to a specific level of energy. For example, long-wavelength radio waves carry a low amount of energy compared with short-wavelength, high-energy X-rays. The entire breadth of the Sun's radiation and range of wavelengths is called the *electromagnetic spectrum,* and science often refers to electromagnetic waves as rays, such as cosmic rays. The following table describes the Sun's electromagnetic spectrum.

THE ELECTROMAGNETIC SPECTRUM		
ELECTROMAGNETIC WAVE TYPE	APPROXIMATE WAVELENGTH RANGE (M)	GENERAL ENERGY CONTENT
cosmic	$<10^{-14}$	very high
gamma	10^{-14} to 10^{-12}	high
X-rays	10^{-12} to 10^{-8}	high
far ultraviolet	10^{-8} to 10^{-7}	high
near ultraviolet	10^{-7} to 10^{-6}	moderately high
visible light	10^{-6} to 10^{-5}	moderate
near infrared	10^{-5}	moderately low
far infrared	10^{-5} to 10^{-3}	low
microwave	10^{-3} to 10^{-2}	low
television	10^{-2} to 10^{-1}	very low
radio	1	very low

The Sun's electromagnetic radiation originates from nuclear fusion reactions in which enormous amounts of hydrogen gas break apart to form helium and energy. Up to 99 percent of the hydrogen combines to form helium molecules and only 1 percent of the fusion reactions produce energy available to the solar system. That small percentage of the Sun's total energy nonetheless represents a tremendous amount of energy. The Sun produces 386^{33} *ergs* of energy per second equivalent to 386×10^{18} *megawatts*. To put these units of energy into perspective, the explosion of 2.2 pounds (1 kg) of TNT releases 1 megawatt of energy.

The Sun emits energy as gamma rays, which travel outward into space. As the gamma rays travel toward Earth, they lose energy in the form of heat. By the time the Sun's radiation reaches Earth, the gamma rays have been transformed to radiation mainly in the visible range of light, meaning light that people can see. Photosynthetic organisms—plant life and some microbes—capture the Sun's radiant energy, called solar energy, when sunlight hits the Earth's surface. The Earth stores solar energy in chemical bonds produced during photosynthesis. Plants use part of this energy. Animals that eat the plants or photosynthetic microbes then receive their energy. When larger animals prey on smaller animals, the predators get a portion of the solar energy and so on until solar energy transfers up an entire food chain. Animals use the energy for moving, breathing, thinking, and functions that keep them alive. At each point in which the Sun's energy transfers from one type of living thing to another, a small amount of energy dissipates as heat. This gradual loss of the Sun's energy follows the second law of thermodynamics, which states that some energy is lost whenever energy changes from one type to another.

A person, a plant, or a microbe cannot transfer solar energy as a ball of light, so living things use another type of currency to move energy from organism to organism. The element carbon serves as this currency. Photosynthesis builds carbon-containing compounds, called organic compounds, to store energy. When animals eat plants or other animals, they get most of the energy they need from organic compounds.

CARBON ECONOMICS

Carbon is the sixth most abundant element on Earth but represents only 0.09 percent of the mass of the Earth's crust. Carbon occurs naturally in all living cells and is a main component in proteins, fats, carbohydrates, nucleic acids (deoxyribonucleic acid [DNA] and ribonucleic acid [RNA]), and vitamins. In fact, of all life-sustaining compounds on Earth, only mineral salts, water, and oxygen gas lack carbon. The Earth's fossil fuels—coal, oil, and natural gas—also contain carbon as their main element because they originally came from living things millions of years ago. On the Earth's surface, forests, the ocean, and fossil fuels act as the main carbon stores.

Humans are like any other living organism; they cannot exist without carbon. Carbon therefore has become a valuable commodity in society. However, it also brings two harms to the environment. First, carbon in the form of the gases carbon dioxide (CO_2) or methane (CH_4) makes up part of the atmosphere's greenhouse gases. Greenhouse gases have aided life throughout the Earth's history by holding in the Sun's heat energy and making the planet a temperate place for life to evolve. Since about 1900, however, greenhouse gases have built up in the atmosphere and caused average global temperatures to rise. Second, carbon compounds given off in the combustion of fossil fuels combine with other elements in the atmosphere to form acids that lead to acid rain. Acid rain has had very damaging effects on plant and tree health and the normal chemical conditions in the ocean. The Earth's carbon therefore presents people with a dilemma: People need carbon as a nutrient, but they must take care to manage the use of carbon compounds to avoid harming the planet.

Carbon economics represents a manner of keeping track of beneficial forms of carbon—as an energy storage material—and harmful forms of carbon—greenhouse gases. Carbon economics consists of buying or selling carbon units, called carbon offsets, on a world trading market, similar to how stocks are bought and sold on the New York Stock Exchange. In North America, businesses conduct carbon transactions on the Chicago Climate Exchange, which opened in 2003. The University of Chicago economist Ronald Coase developed the concept of trading carbon units in the 1960s. Coase's journal article "The Problem of Social Cost" examined the relationships between business actions and the well-being of communities in the context of the environment: "The standard example is that of a factory the smoke from which has harmful effects on those occupying neighboring properties." Moving the factory or shutting it down takes jobs from the community. A community may therefore decide that it is willing to endure the disadvantages for the advantage of keeping its livelihood.

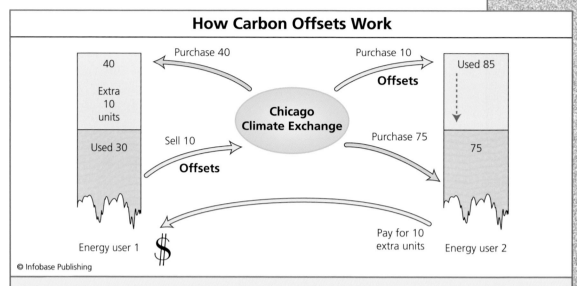

How Carbon Offsets Work

Purchase 40   Purchase 10

40

Extra
10
units

Used 30   Sell 10

Offsets

Chicago
Climate Exchange

Offsets

Used 85

Purchase 75

75

Pay for 10
extra units

Energy user 1 $

Energy user 2

© Infobase Publishing

Carbon or emissions trading, the buying and selling of carbon, has been praised as one of the most innovative methods for getting industries to lower their greenhouse gas emissions. The Chicago Climate Exchange serves as the main trading house in the United States. Other climate exchanges operate in Europe (the largest climate exchange), Canada, Australia, China, and Japan, with several new exchanges due to open. Climate experts have not yet found evidence that the exchanges have affected global warming.

Coase wrote, "We are dealing with a problem of a reciprocal nature." In other words, people sometimes take reciprocal actions. This means that they might sometimes choose a harmful course of action if the gains of that action outweigh the losses.

Carbon economics tries to achieve greater gains while lessening the ill effects of carbon in the world. The following table presents the main aspects of carbon economics.

Carbon trading plays a central role in carbon economics, but many people question whether carbon trading actually contributes to pollution and global warming. A person or company that stays below its allowable limit of emissions may sell extra carbon units to companies that have exceeded their emissions target. Critics of carbon trading say that the plan simply allows polluters to continue polluting as long as they are willing to pay a fee. Carbon trading rewards businesses that produce low emissions by making it possible to earn extra money by selling carbon credits. Carbon trading also gives polluters extra time to achieve emissions limits that will become stricter over time.

CARBON ECONOMICS	
CARBON TRANSACTION	**DESCRIPTION**
carbon dioxide equivalent (CO_2e)	a measure used to indicate the global warming potential of a gas emission relative to CO_2
credits	amount of CO_2e that a business can sell if that business conducts activities known to limit greenhouse gas emissions
markets	places or institutions that bring buyers and sellers of carbon credits together
carbon trading (also emission trading)	the scheme whereby companies sell CO_2e to polluters (also called carbon offsetters)
offsets	a unit of CO_2e that can be purchased by a polluter to be applied against that polluter's excess emissions (often used interchangeably with carbon credits)
domestic tradable quota	the entire process of buying and selling CO_2e for the purpose of rationing the use of fossil fuels and thereby lowering greenhouse gas emissions
tax	a tax levied on any polluter that exceeds its legal limit of emissions, based on the amount of emissions over the limit
direct payment	payment from a governing agency to any business that produces less emissions than its allowable limit, based on the amount of emissions under the limit
cap and trade	system in which a limit is set on the amount of emissions allowed by a business—the cap—which, if exceeded, the business must buy offsets on the carbon trading market
assigned amount unit	a tradable unit of CO_2 in the form of 1 ton (0.91 metric ton) of CO_2e

The international treaty called the *Kyoto Protocol* has backed carbon trading as a benefit to the environment, and according to the Chicago Climate Exchange, "Application of flexible, market-based mechanisms for reducing greenhouse gas emissions has achieved widespread intellectual and political support. This broad acceptance of emissions trading was reflected in the Kyoto Protocol, which established several emissions trading mechanisms." Though the Chicago Climate Exchange has stated that carbon trading "makes good business sense and environmental sense," others disagree. Carbon market analyst Veronique Bugnion said in the *San Francisco Chronicle* in 2007, "Have they [carbon markets] achieved any real reductions in greenhouse gases? There is not much evidence of a reduction." It may be too soon to tell if carbon trading can slow global warming, but the World Bank has predicted that carbon trading will soon become the largest commodities market in the world. From 2005 to 2006, the global carbon market more than doubled the amount of carbon equivalents that moved between buyers and sellers. This amount rose 63 percent in 2007 and 83 percent in 2008. In 2008 alone, 5.4 billion tons (4.9 billion metric tons) of carbon equivalents had changed hands.

OUR RENEWABLE ENERGY RESOURCES

Switching from fossil fuel burning for energy production to renewable energy sources lowers the total amount of carbon released into the atmosphere as CO_2 gas. Six main types of renewable energies have been employed in industrialized places for this purpose and are listed in the following table. As the table shows, renewable technologies may be either modern advances in energy generation or ancient technologies that some parts of the world continue to use. Solar, water, and wind energy plus the burning of organic wastes together account for 7 percent of energy consumption in the United States and about 20 percent worldwide. Fossil fuels and nuclear power supply the rest.

Of the main types of renewable energy, only biomass puts CO_2 into the atmosphere. Burning biomass offers a good environmental choice only if the rate of burning biomass does not exceed the rate of new plant growth on Earth. Put another way, plants must be able to remove more CO_2 from the atmosphere than burning puts into the atmosphere.

Many renewable energy sources do not produce usable energy directly, and equipment must convert one type of energy into another form. For

RENEWABLE FORMS OF ENERGY			
ENERGY SOURCE	PERCENT OF RENEWABLE SOURCES	DESCRIPTION	PRODUCT
biomass	53	burning of plant materials and animal wastes	heat and gas
hydropower	36	water flowing from higher to lower elevations through dams	electricity
wind	5	capture of wind by *turbines*	electricity
geothermal	5	tapping steam and hot water from the Earth's mantle	heat and electricity
solar	1	absorbing and storing heat from the Sun	heat and electricity
emerging technologies			
hydrogen fuel		burning hydrogen gas	power for movement
nanotechnology		using the unique properties of materials on the size scale of molecules or atoms	electricity
ancient technologies			
water		water wheels, dams, weight	power, motion
wind		windmills, sails	power, motion
movement (*kinetic energy*)		animals, human exertion	power, motion

example, the energy contained in wind turns a turbine, which powers a generator that makes electricity. Energy contained in motion, such as wind or flowing water, is kinetic energy. Sometimes kinetic energy helps convert one form to another, such as the wind turbine mentioned here, or kinetic energy itself might be used. An ox pulling a plow across a field is an example of kinetic energy at work.

SMART ENERGY GRIDS

An energy grid or a power grid consists of a large distribution network that carries electricity or natural gas from producers to customers. The United States contains a large electrical power grid, but smaller regional grids also supply electricity to customers in, for example, southwestern states. The U.S. natural gas grid consists of thousands of miles of underground pipes that distribute natural gas throughout the contiguous 48 states.

Conventional energy grids that have been used for many years distribute energy in a one-way fashion. A large power plant generates electricity, which enters high-voltage power lines that take the electricity to smaller, local electric utilities. Hydroelectric dams and *coal-fired power plants* supply most of the electricity on this type of energy grid. The final consumer—houses or businesses—then draws electricity off the grid and pays for the amount taken. This system has been convenient for customers, but it has also been wasteful. Even though customers pay for the electricity they take, a significant amount of energy is wasted when people do not turn off or unplug electronic devices when not in use. On the other end of the grid, coal-fired plants generate a troubling amount of emissions even if they are equipped with emission-reducing devices such as *scrubbers*. Hydroelectric dams also receive blame for harming *riparian ecosystems* by releasing hot *process water* into the environment.

Smart energy grids improve on conventional energy distribution in two ways. First, smart energy grids maximize the use of alternative energy sources that supply electricity without causing the environmental harm associated with dams and coal burning. Large power plants and thousands of miles of power lines can be eliminated by the use of smart energy grids. Second, smart energy grids can be designed to eliminate or minimize waste by allowing a type of two-way flow of electricity. These grids make it easier for consumers to pay for what they use rather than what they take.

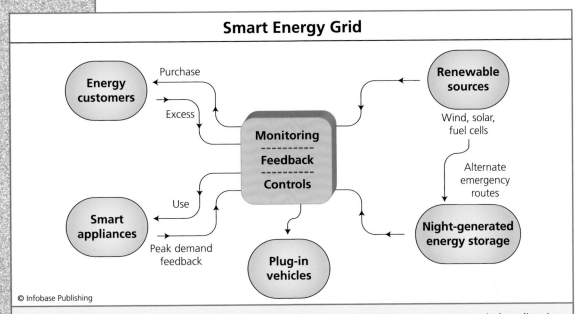

Smart Energy Grid

Energy customers

Purchase

Excess

Renewable sources

Wind, solar, fuel cells

Monitoring

Feedback

Controls

Alternate emergency routes

Smart appliances

Use

Peak demand feedback

Plug-in vehicles

Night-generated energy storage

© Infobase Publishing

Smart grids will contain the following: feedback systems to monitor peak and low energy usage periods, redirection of extra energy to places with a temporary energy demand, devices for storing wind and solar energy, alternate emergency routes to avoid system breakdowns, and accommodations for smart appliances, electric vehicles, and other new technologies as they become available.

A smart energy grid is made up of two main components: a power plant and a computerized system that constantly monitors electricity usage. Future smart grids will likely use renewable energy sources. The computerized monitoring system keeps tabs on the times and locations of highest electricity use and can redirect power at any moment from low-use locations to high-use locations. Advanced smart grids may soon connect to in-home smart appliances, which also sense peak usage times. The appliances send information to the grid to indicate a lesser or greater need for electricity. This two-way communication between the consumer and the energy grid is called *feedback* and is the key element in regulating energy use in a more responsible manner. In times of high and sudden power usage, such as evenings when computers, kitchen appliances, and heating or cooling systems all run at once, smart grids parse out energy in a staggered fashion to reduce strain and waste on the entire system.

Lou Schwartz and Ryan Hodum, writing for RenewableEnergyWorld. com in 2008, explained, "In the United States . . . although the transmission

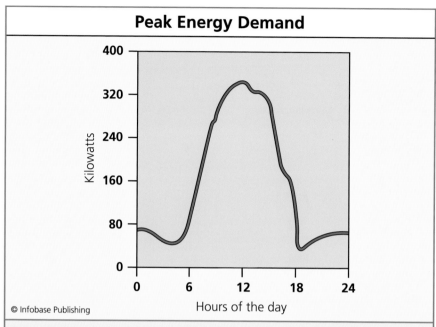

Peak Energy Demand

© Infobase Publishing

Hours of the day

Engineers design smart energy grids by studying the peak energy use periods of the grid's customers. In the United States, most peak energy patterns follow the graph shown here. Energy use falls during the night and peaks in the middle part of the day. More in-depth studies would show that different activities have differing energy usage patterns. For example, a large factory working three shifts has a different pattern than a small office building.

grid is 99.97 percent reliable, brief power interruptions have cost the country nearly $100 billion each year; apart from enhanced reliability, smart grids promise to increase efficiency of power distribution and usage, with corresponding savings in power and power consumption." China currently is planning to overhaul its electricity supply by adopting smart energy grids, and Europe and Australia have also made investments in smart systems.

SOCIAL ASPECTS OF ALTERNATIVE ENERGY

Homeowners and businesses often cite the expense of new alternative energy sources as a reason for staying with conventional energy supplied by power plants. Solar energy indeed carries a large price tag, and in some cases the money saved on smaller electric bills does not pay for the system's purchase price and installation until decades later. For this reason, many people regard alternative energies such as solar a luxury item.

In impoverished parts of the world, millions of people are starving, and they are not worried about efficient energy supply. But sustainable use of energy in developing parts of the world encourages the use of local resources, reduces health-threatening pollution, and creates jobs. No one has proven that sustainable energy can help alleviate poverty. Sustainable practices do, however, make people more aware of their environment and its potential destruction. Developing regions furthermore do not have to undo habits associated with the industrialized world, such as the use of numerous electronic devices, large energy-demanding homes, and high-maintenance luxuries (swimming pools, sports cars, video game consoles, etc.). Because of this, industrialized nations and international organizations can help developing parts of the world build sustainability from the ground up.

Countries that have been mired in poverty for generations and which now are taking bold steps toward industrialization have gone from being energy-efficient places to high energy-consuming places. The IEA expects China and India to account for more than half of world energy demand approaching the year 2030. Both countries rely on oil and coal for power, and their power plants produce enormous pollution. BBC correspondent James Reynolds described a situation in China's Shanxi Province: "At a temperature of –10°C (14°F), in the grey-blue dawn, two schoolchildren have a thankless job to complete. They are meant to sweep away the soot, dirt and grime from the school gate. But this village is surrounded by coal mines and power stations, so it is impossible to get anything clean." Their environment certainly would have looked better if these countries had built their industrial revolutions based on sustainable fuels rather than nonrenewable and polluting fossil fuels.

Some countries that have been building strong industrialized economies—China, India, parts of the Middle East, South Korea—have developed existing conventional energies rather than new technologies. Their leaders may emphasize commerce over environmental protection. When a large country or small community decides on the types of energy source it will use, the decision encompasses more than blueprints for power plants and power lines. Energy decisions should be based on each region's economy, its local resources, and the willingness and ability of leaders to work with citizens in making the correct energy choices. The U.S. federal government has developed several agencies that monitor energy law, evaluate new technologies, and guide the populace on how to make sound energy conservation choices. The sidebar "The U.S. House Subcommittee on Energy and Environment" on page 28 discusses one such organization.

ENERGY PROGRAMS IN THE GLOBAL COMMUNITY

The Global Energy Network Institute has been working since 1986 to help design a global system for distributing energy. As part of the San Diego institute's plan, the new network would distribute power derived mainly from renewable energy sources. The scientist R. Buckminster Fuller laid out his vision for the global energy network: "Electrical-energy integration of the night and day regions of the Earth will bring all the capacity into use at all times, thus overnight doubling the generating capacity of humanity because it will integrate all the most extreme night-day peaks and valleys." The global network in this way has the potential of moving energy around

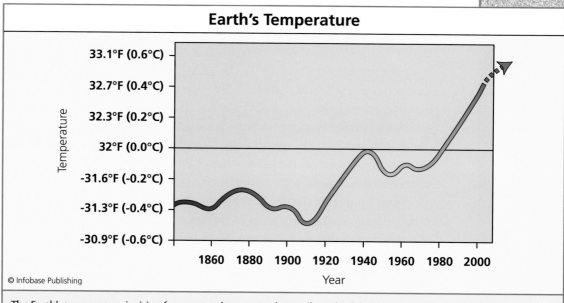

Earth's Temperature

© Infobase Publishing

The Earth's temperature is rising for reasons that cannot be attributed solely to natural cycles. Human activities that have increased with population growth have caused a rise in the globe's overall average temperature. Some of the many methods that combine to produce enormous databases on temperatures around the world are weather station readings, satellite ocean and landmass measurements, ocean and atmospheric readings, polar ice permeability, and polar ice and glacier melt rates.

the world to places where energy is needed most. Fuller's far-reaching plan requires large investments in money and time plus commitment among nations. Surely a global energy grid would present large obstacles along the way to success.

A global energy network requires that leaders gather energy experts to work on specific phases of building such a network. The IEA, for instance, has representatives from 20 industrialized countries working on current energy questions. Each year, the IEA presents an update on the world's energy usage and production, regions with overconsumption, and offers ideas for future energy management. The following list contains the IEA's most recent observations on global energy:

- All current energy trends are not sustainable.
- Oil will likely remain the leading energy source.
- Oil fields have declined, so undiscovered oil reserves will be needed to sustain current consumption.
- Countries must cooperate on holding the average global temperature rise to 3.6°F (2°C).
- Arresting current global temperature rise requires lowering emissions in both industrialized and nonindustrialized regions.

The world contains diverse economies and customs, so building a global energy program will be very difficult. For example, a standard energy grid cannot meet New York City's needs in the same way it would serve Mongolia. IEA's director Nobuo Tanaka stated in a 2008 press release, "We cannot let the financial and economic crisis [of 2008] delay the policy action that is urgently needed to ensure secure energy supplies and to curtail rising emissions of greenhouse gases. We must usher in a global energy revolution by improving energy efficiency and increasing the deployment of low-carbon energy." By "low-carbon energy," Tanaka means solar, wind, water, and nuclear sources in place of burning fossil fuels or wood. The IEA and other international organizations put considerable efforts into balancing the ways to minimize energy consumption, stop environmental decay, and address social issues such as poverty. Appendix C lists the major international groups with active energy policy programs and plans for reaching sustainability.

THE U.S. HOUSE SUBCOMMITTEE ON ENERGY AND ENVIRONMENT

The U.S. House of Representatives Committee on Science and Technology has five subcommittees: space and aeronautics; technology and innovation; research and science education; investigations and oversight; and energy and environment. The Committee on Science and Technology was formed by Congress in 1958 in response to the Soviet Union's launching of the *Sputnik* spacecraft the year before. Feeling that the United States might be falling behind in the technology race, Congress asked for increased programs in science and innovative technologies. The BBC reported at the time of the launch, "There have already been calls for an immediate review of U.S. defenses, given the implications of the technological leap ahead by a political enemy." New studies would be designed not only for space flight, but also physics, weapons, and environmental studies.

In the 1960s and 1970s, the public turned its attention increasingly to the environment. Air and water pollution, hazardous wastes, environmental accidents, and the loss of species and their habitats seemed to affect every region. In literature, ecologist Rachel Carson alerted readers to the dangers of pesticides in her 1962 book *Silent Spring*, and a 1966 science fiction novel by Harry Harrison, *Make Room! Make Room!*, provided the story line for a 1973 movie *Soylent Green*, which popularized the term *greenhouse effect*. People were becoming aware of the real problems of gas emissions, chemical-leaking dumpsites, and wastes pouring into the ocean. The environmental subcommittee began to gain an audience for developing new ideas for repairing environmental harm.

By the 1980s, the federal government was tackling mounting problems in waste disposal, environmental hazard, pollution controls, environmental health, and new energy sources. After the 1994 national election, the U.S. House of Representatives reorganized the Committee on Science and Technology into subcommittees to cover specific focus areas. Dana Rohrabacher of California became the chairman of the new Subcommittee on Energy and Environment. Today,

CONCLUSION

The Earth's energy sources have been sufficient for sustaining human life since its inception. Humanity in the current era depends mainly on fossil fuels—petroleum, natural gas, and coal—to keep industrialized and unindustrialized countries running. This plan has worked for generations, but since the 1970s and 1980s scientists have sent out alarms regarding the human population's insatiable appetite for energy. Some scholars have calculated that people are nearing a point at which more than half of

the chair is held by Brian Baird of Washington. The subcommittee continues its role in evaluating new approaches to energy use and conducting hearings with experts on the effects of fossil fuel use on pollution and global warming.

The Subcommittee on Energy and Environment's jurisdiction has now expanded to the following main areas:

- Department of Energy research, laboratories, and other science activities
- renewable energy technologies
- nuclear power materials, wastes, and safety
- fossil fuel energy and pipeline research
- alternative energy sources
- energy conservation
- National Oceanic and Atmospheric Administration (NOAA) activities in weather, climate, and ocean conditions

The subcommittee encourages academic and government researchers to move quickly in developing the areas listed above. These topics are no longer projects for the future; they have all now become urgent. Although government has been accused of moving too slowly in times of crisis, environmental scientists have shown that climate change and natural resource depletion are true crises without the luxury of time. Organizations such as the Subcommittee on Energy and Environment can help the United States make the right decisions about building sustainability in energy use and natural resources.

petroleum has been used up. Though natural gas and coal seem to remain plentiful, no rational person can deny that they too will reach an end.

The concept of sustainability relates to resources in addition to energy: land, clean water, clean air, and the tremendous species diversity. Energy sustainability focuses on technologies to slow the rate at which people devour fossil fuels. This slowing will come about only if two things happen. First, new technologies in renewable energy must replace fossil fuels as the main energy source. Second, people must make serious attempts to conserve all energy use from both nonrenewable and renewable sources.

Energy overconsumption certainly derives from mismanagement of energy so that energy waste takes place. Overconsumption also comes from the desire to own more things than needed. Excessive use of energy resources also relates to the growth rate of the human population. Even if every person adopted a lifestyle that conserved energy, the population would simply overrun the Earth's capacity to sustain it in many places. This situation, defined by the ecological footprint, indicates that the world can no longer conduct business as usual regarding fuel and electricity consumption.

Government leaders have suggested that the solution to an energy debt is to find more hidden fossil fuels in the Earth. Environmentalists counter that such exploration does not solve the energy problem and, in fact, leads to more pollution from burning more fossil fuels. Renewable energy sources from the Sun, water, and wind offer a more sustainable future than the dependence on fossil fuels. Renewable energies have obstacles to overcome to be sure, but none of the obstacles appear to be outside of mankind's abilities. Perhaps the next generation will know that renewable energy has arrived when the use of nonrenewable fuels seems obsolete.

Renewable energies have no single road to success. Like the telecommunications industry and the computer industry before it, energy technology's advances will probably come from different approaches to meeting a need. Communities dependent on renewable energy will likely use a combination of solar, wind, hydroelectric, and even nuclear energy. This differs little from the way in which countries today use coal, natural gas, and oil. The new and growing renewable energy industry will attack the energy problem from many fronts in order to meet human needs. Everyone should also remember that sustainability, however successful, does not last forever. Today's innovations in renewable energy prolong sustainability. Innovations that have barely been imagined will be needed to affect sustainability 100 years from now and beyond.

RECYCLING

The action of recycling wastes seems to symbolize the idea of sustainability. Recycling can be done by a person living in a small apartment or by a massive factory. Recycling fulfills two components of sustainability. First, people conserve natural resources by recycling items that industries use as raw materials. This decreases the demand that industry puts on the environment by extracting new natural resources from the Earth. Second, recycling lessens the amount of wastes that accumulate on Earth. The simple action of putting wastes into different recycling bins also helps remind people of the amounts of waste they produce and might help them think of ways to reduce it.

Most communities in the United States have recycling programs in which families, businesses, and schools participate. Some critics of recycling, however, have pointed out that ambitious recycling programs, however well-intentioned, do not help the environment. These critics feel recycling uses more energy than it takes to make products directly from natural resources. In 1996, the *New York Times* writer John Tierney initiated the recycling debate in his article "Recycling Is Garbage." Tierney stated that landfill space was abundant—despite evidence to the contrary—and that landfills offered a wiser approach to waste disposal than recycling. "Recycling may be the most wasteful activity in modern America," he wrote, "a waste of time and money, a waste of human and natural resources." Improvements in recycling technology since 1996 have made recycling a viable alternative to landfills and a means of building sustainable use of resources.

The National Recycling Coalition (NRC) countered Tierney's opinion with data that show how recycling saves energy compared with making products from new raw materials. The following recycled materials save energy compared with manufacturing them:

- aluminum, 95 percent
- plastics, 70 percent
- steel, 60 percent
- newspaper, 40 percent
- glass, 40 percent

The NRC further advises all communities to concentrate on the following 10 items for recycling, in order of importance to industry, to save the most energy and resources overall in the environment: aluminum, polyethylene terephthalate (PET) plastic bottles, newspaper, corrugated cardboard, steel cans, high-density polyethylene (HDPE) plastic bottles, glass containers, magazines, mixed paper, and computers. The following table shows that despite the growth of recycling programs, the United States has room for improvement in recycling.

RECOVERY OF RECYCLABLE MATERIALS IN THE UNITED STATES		
MATERIAL	WEIGHT GENERATED IN MILLION TONS (MILLION METRIC TONS)	APPROXIMATE PERCENT RECOVERED FOR RECYCLING
paper	86 (78)	50
plastics	29 (26)	11
glass	15 (13.6)	100
steel	14 (12.7)	48
aluminum	4 (3.6)	30
Source: Greenstar North America		

Recycling will not solve all environmental ills. To achieve sustainability, people must do more than recycle to conserve natural resources. But recycling certainly helps lessen pollution, waste, and natural resource depletion, even if it alone cannot fix these problems. Recycling technology continues to grow, and entrepreneurs have invented new uses for wastes while the recycling industry has found ways to make recycling less expensive and more streamlined.

This chapter reviews the history of U.S. recycling programs and looks into methods in which technology has improved energy savings. The chapter discusses specialties in the recycling industry such as metals and rubber recycling. It also reviews the chemistry involved in turning a waste material into a recycled material. In addition, this chapter provides an example of one of history's largest recycling programs, which took place during World War II. Well-managed recycling programs have contributed and will likely continue to aid in sustainability.

THE GRASSROOTS HISTORY OF RECYCLING

Recycling has been part of civilization for thousands of years. In 1030 B.C.E., Japan employed an organized system of collecting wastepaper for the purpose of turning it into new recycled paper. Little recycling or waste management seems to have taken place during the Middle Ages. Recycling returned, however, as a way to make businesses more profitable. In 1690, the Rittenhouse Mill near Philadelphia turned rags from used cotton or linen into new paper. England and the new colonies established a variety of recycling businesses from that point onward, reusing metals, paper, and cloth. By the mid-1800s in the United States, peddlers who traveled door to door paid a few pennies to families in return for any discarded items. The peddlers then resold the items to craftsmen. By the end of the century, some towns had set up recycling programs similar to the curbside pickup programs used today—the first curbside program began in Baltimore in 1875.

Recyclers carved out enterprises in large cities in the early 20th century, putting aluminum cans, twine, rubber, and burlap bags to new uses. Cities such as New York built organized recycling programs; Chicago put its prisoners to work sorting waste. World War I and II increased the necessity to salvage as much recyclable material as possible. For this purpose, the federal government set up the Waste Reclamation Service during World War I to run a recycling effort. In World War II, the War Production Board's Salvage Division ran one of the most ambitious recycling programs ever established.

Prosperity returned in the years following World War II and with it came a variety of convenience products that encouraged disposal rather than reuse. By the 1960s, hazardous wastes created serious health threats on land and in water. The public and the U.S. Congress began to see waste as a serious national problem, and by 1965 Congress passed the Solid Waste Disposal Act to assist local governments in setting up waste programs. The aluminum industry took the lead by building a large project for recovering and reprocessing beverage cans. Little by little, towns built centralized recycling centers to help take in aluminum and paper. The idea blossomed, and within two decades the United States had 10,000 recycling centers.

The first Earth Day on April 22, 1970, signified a shift in the relationship between the public and the environment. Communities, families, and students became committed to environmental projects. Waste, pollution, habitat loss, and biodiversity grew into topics of conversation among university professors as well as the public. A new type of environmentalism called *grassroots* environmentalism began. Community and school recycling leaders reminded anyone who would listen that individuals could come together for a single purpose and make a difference in aiding the environment.

In 1989, the University of Arizona archaeologist William Rathje led his students on an assignment they called the "Garbage Project." The team set out to investigate landfills to learn about how Americans produced and discarded waste. Rathje noted what their collections revealed: "Despite all of the concern directed at fast food packaging and disposable diapers, the archaeological data demonstrated that both items together accounted for less than 2 percent of landfill items. . . . By volume nearly half of all of the refuse excavated by the Garbage Project has been newspapers, magazines, packaging paper and non-packaging paper, such as computer printouts and phonebooks." Rathje also found large volumes of construction and demolition wastes, adding to a vast amount of resources available for recycling but going to waste.

The American public embraced certain aspects of recycling with more enthusiasm than others. For example, by 1995 Americans had recycled more than 47.5 billion aluminum containers, but they did a poor job in recycling paper (as the Garbage Project showed). Some communities took recycling more seriously than others. Many California towns embraced grassroots recycling projects with such enthusiasm that the state government took notice and adopted many of the local processes. Nationwide, an environmental organization called the Grassroots Recycling Network became a resource for communities that wanted to start their own recycling programs. Today, recycling advocates work toward the achievement of *zero waste* in which near 100 percent of all wastes can be put to use. The Zero Waste Alliance of Portland, Oregon, has explained, "Zero waste strategies consider the entire life-cycles of our products . . . With this understanding, wastes can be prevented through designs based on full life-cycle thinking. Indeed, we should work to 'design' our wastes, if any, so that they have future applications." Until society achieves success in zero waste, recycling serves an important role in natural resource conservation.

HOW RECYCLING SAVES ENERGY

Two types of recycling exist for the purpose of conserving natural resources and, whenever possible, saving energy. *Primary recycling,* also called closed-loop recycling, turns recycled materials into new products of the same type. For example, used aluminum beverage cans are recycled into new beverage cans. *Secondary recycling,* also called downcycling, recycles materials into new and different products, such as used plastic milk jugs used for new decking and outdoor furniture. Neither type of recycling would succeed if the costs of recycling a material exceeded the costs of making the product out of new raw materials. Even if the difference in costs is small between a recycled product and a new product, recycling helps the overall finances of a community by reducing the amount of waste that must be incinerated, put in a landfill, or otherwise treated.

Industries strive to use recycling processes that demand less energy than manufacturing a new product from raw materials. Facilities called materials recovery centers also help hold down costs by doing some of the work for consumers and manufacturers. Once a container of recyclable materials has been delivered to the center, either by a resident or a commercial waste hauler, the recovery center carries out the following steps:

Recycling begins with separating and sorting waste types to recover all the materials that can be recycled. Sorting plays an important role in recycling because small amounts of contamination, plastic mixed in with aluminum for example, decreases the recycling efficiency, which is critical to keep the recycling process cost-effective. These sorters in an Oregon recycling plant are removing non-paper materials from paper wastes. *(OregonLive.com)*

- sorting—nonrecyclable from recyclable materials and hazardous from nonhazardous materials

- separating—types of paper, plastics, glass, and metals, such as brown glass from green glass bottles

- treatment—sending nonrecyclable materials to a final disposal site, such as an incinerator or a landfill

- recovery—sending materials to a business that uses them as raw material, such as steel sent to automakers

The steps shown here usually consume less energy than the steps needed to make a product from new raw materials. For many recycled materials, the sorting, processing, and transportation use less energy than the following steps needed for making new raw materials: (1) exploration, (2) extraction, (3) transportation, (4) processing, and (5) waste treatment. Aluminum recycling may be the best example of how this works. A recycled aluminum can requires only 5 percent of the energy that would be needed to make the same can from virgin (new) aluminum. The Stop Global Warming Organization based in California has noted, "Using 1 ton [0.9 metric ton] of recycled aluminum as opposed to 1 ton of virgin aluminum would power an American household for over 15 months." Recyclers would like to find the same efficiencies in other materials such as glass, paper, and plastic.

The recycling balance sheet for some materials does not always work as well as that for aluminum for two reasons. First, some recycling steps cost more than new materials. Second, sometimes recycling simply cannot keep up with the amount of recyclable waste that comes into recycling centers. When recycling cannot keep up with waste production, recyclable items accumulate. Jim Hogan, manager of a recycling center in Yonkers, New York, told the Lower Hudson Valley *Journal News* in 2006, "Whatever we can take out of the waste stream is a plus. Crushed glass is [costing us] well over $100,000 a year. Most of it gets used as landfill cover . . ." The potential to save energy in recycling the world's wastes, therefore, is connected to recycling costs and savings.

Recycling saves energy and money only if it meets two requirements. First, a sufficient amount of material must go into the recycling process to make recycling both energy- and cost-efficient. Large operations usually cost less per unit, in energy and in money, than small processes. This phenomenon is called *economy of scale,* in which companies receive advantages by using materials in bulk and producing large quantities of product in a single production run. For example, a wristwatch made by a company that produces thousands of watches a year costs less to make than a watch built by a single watchmaker working in a small shop. Second, there must be customers for a recycled product. High demand for a recycled material allows the recycler or manufacturer to take advantage of economies of scale. Consumers thus keep the entire recycling loop going in an efficient manner.

The need for very high efficiency in order to make recycling worthwhile has caused some people to criticize curbside recycling programs. In 1996, the *New York Times* columnist John Tierney wrote, "Believing that there was no more room in landfills, Americans concluded that recycling was their only option. . . . Recycling does sometimes make sense—for some materials in some places at some times. But the simplest and cheapest option is usually to bury garbage in an environmentally safe landfill."

Tierney's conclusion works only if, as he believed, landfill space is plentiful. Landfill space in most states has been expanded, yet waste managers have calculated that the last space will be used up within 20 years.

Others in addition to Tierney have expressed varied opinions on recycling. The *On Milwaukee* magazine editor Drew Olson explained in 2008, "Anti-recyclers claim that gains derived from curbside pickup are offset by the need for more trucks, which consume more gas and create more atmospheric pollution. They feel that recycling advocates, aided by the media, created a number of myths about recycling." In truth, not all recycling leads to the advantages enjoyed by the aluminum industry. Most recyclers and manufacturers must watch their expenses and energy use very carefully to make recycling work in their favor.

Plastic recycling presents more obstacles than aluminum recycling. In 2007, *Science Daily* noted, "While major cities like New York and San Francisco have shown that plastics recycling can be done successfully on a large scale . . . many municipalities are still falling far short of their recycling goals." Handled correctly, the U.S. Environmental Protection Agency (EPA) states that recycling aluminum saves 95 percent of the energy of virgin aluminum, recycling paper saves 60 percent, but recycling glass saves

The recycling industry has made important advances in increased speed and efficiency of recycling and in finding new uses—in business, this is called developing new markets—for more materials. These bales of recycled materials have been separated from other wastes and perhaps sorted into additional categories. Baling makes handling and transport easier, and overall this step will likely save fuel and money. *(South Dakota Department of Environment and Natural Resources)*

less than 50 percent. For plastics, the efficiencies of recycling depend on the type of compounds in the plastic. The greatest needs in recycling technology are improved efficiencies in glass and plastic recycling so that these methods save energy and money.

INDUSTRIAL RAW MATERIALS FROM WASTE

Recycling efficiency has improved because of advanced materials science and chemistry. Materials scientists and chemists study the behavior of recyclable waste when crushed, heated, extruded, or otherwise made into new product. Schools such as the Massachusetts Institute of Technology have combined both disciplines into a new area called materials chemistry. Materials chemistry encompasses the following subject areas that can be applied to recycling: organic and inorganic chemistry, physical chemistry, *polymers,* biochemistry, and analytical chemistry.

INNOVATIVE RECYCLING	
WASTE MATERIAL	NEW USE
bamboo	microbe-resistant workout clothes
coconut	odor-absorbing clothes
denim fiber	insulation for houses
inner tubes	purses and handbags
paper	cat litter
plastic	outdoor blankets
soybean husks	baby clothes and blankets
tires	sealants, shoe soles, paint
tubs, sinks, and toilets	terrazzo counters and floors

INDUSTRIAL RECYCLING		
MATERIAL	**SOURCE**	**INDUSTRIAL USE**
coal fly ash	inorganic material left over from the combustion of coal	mixed with concrete for retaining walls, pavement, and structural fill
construction and demolition debris	concrete, bricks, steel, sheet metal, roofing, and wood	crushed and mixed with asphalt as pavement
foundry sand	excess processing material used in metal casting	structural fill in construction; cement manufacture; landscaping topsoil; grout; mortar
gypsum (a soft mineral found in nature)	plasterboard waste	new plasterboard, construction materials
slag (excess runoff during metal manufacturing)	heterogeneous by-product of ore smelting	metal manufacturing

As a mark of how successful recycling has become, the industry now manages more than 1,000 recycled product categories. About 80 different recycled materials serve as the raw materials for making these products. Appendix D lists the important recyclable materials that go into making a wide variety of products sold to consumers today.

Entrepreneurs also play a role in the recycling industry by recovering materials that large recycling businesses cannot use profitably. Many unique uses have come out of these innovative recycling schemes. Examples in the table on page 39 show that small businesses with a good idea on how to reuse a particular waste can contribute to sustainability.

Large industries create great quantities of waste that cost money to treat or may take up space in landfills. Whenever possible, industries send their waste materials to other businesses that can use them as a raw material. Industries refer to this recycling as *beneficial use* because they know their wastes will become a valued commodity for another industry. The table on page 40 describes some of the new uses industries find for wastes.

The plastics industry contends with a more complex recycling process than those described in the preceding table. This is because plastics come in a variety of chemical structures that have their own properties. The sidebar "High-Density Polyethylene (HDPE)" on page 42 describes one of the main plastic components recycled today.

RECYCLED MATERIALS CHEMISTRY

Recycling chemistry draws on aspects of materials science and organic and inorganic chemistry to make recycled wastes suitable for new uses. The chemical aspect of recycling consists of four steps: (1) breaking the material into components; (2) extraction of the target components; (3) cleaning or purification; and (4) analysis. Paper, glass, metal, and plastic recycling require these steps or similar methods as described briefly in the following table. Analysis of the final recycled material takes place either as laboratory chemical tests (plastic resins) or as sensitive measurements on sophisticated instruments (metals).

Recycling chemistry uses common laboratory procedures that have been modified for the particular substance being recovered from an item.

BASIC RECYCLING STEPS			
MATERIAL	**SEPARATION**	**EXTRACTION**	**PURIFICATION**
glass	by color	washing	melting
metal	by metal composition	melting	chemical removal of nonmetals
paper	ink from paper	filtration for recovering cellulose fibers	steam-pressure treatment
plastic	by molecular structure	polymer breakdown to individual units	solvent treatment

HIGH-DENSITY POLYETHYLENE (HDPE)

H DPE is a long carbon-hydrogen compound, called a polymer, formed by applying heat and pressure to the methane gas (CH_4) from petroleum. German chemist Hans von Pechmann developed this method for making the polymer polyethylene in the 1890s. Other chemists followed by changing the types and sizes of side chains attached to the polyethylene. They found that such changes would alter the flexibility of the final polymer. In 1935, the British chemists Eric Fawcett and Reginald Gibson created a durable polyethylene for insulating cables. Almost 20 years later, the German Karl Ziegler devised a chemical reaction to make a dense polyethylene that maintained a rigid and strong structure—the present form of HDPE—and received the 1963 Nobel Prize in chemistry for his discovery. The plastics industry continues to use Ziegler's process to make HDPE from ethylene gas or from natural gas that comes from oil refining.

HDPE is a thermoplastic, meaning it is a plastic in which the molecules are held together by weak bonds so that the material softens when heated but returns to its original condition at room temperature. Thermoplastics are in milk jugs, shampoo or detergent bottles, credit cards, and floor coverings. Thermoset plastics, by contrast, do not lose their shape or strength when heated, and so these plastics work best in vehicle components and construction materials.

HDPE does not degrade in nature and so can quickly take up much space in landfills. (HDPE items total only 1 percent by weight of municipal solid wastes [MSW].) More than 95 percent of all plastic bottles contain HDPE or polyethylene terephthalate (PET), a plastic used mainly for soda bottles. Other materials used to a lesser extent by the plastics industry are low-density polyethylene (LDPE), polypropylene (PP), polystyrene (PS), and polyvinyl chloride (PVC). Each of these plastics has a different content of resin, which is the specific carbon-hydrogen, or hydrocarbon, polymer that makes up the final formula of the plastic. Each resin correlates with a code developed by the Society of the Plastics Industries (SPI) in 1988. The code number

The most common chemical procedures are the following: distillation, filtration, phase separation, and catalyzed reactions. Distillation entails the heating of a material to drive off the water. Distillations of recyclable materials range from room temperature to low-temperature procedures. Filtration involves passing a substance through a barrier containing very small pores (the filter) that allow liquids and small particles to pass through. Chemists save either the material that passes through the filter, called the filtrate, or the material that remains on the filter. Phase separation recov-

usually occurs inside a triangle symbol that has been embossed into the plastic, for instance, on the bottom of a shampoo bottle. The SPI resin codes are the following: PET, 1; HDPE, 2; PVC, 3; LDPE, 4; PP, 5; PS, 6; and all other resin blends, 7.

Recyclers process HDPE first by sorting and washing plastic wastes. Recyclers then chop the plastic into small pieces of less than one-half inch (1.27 cm), called flake. The recycler pours the flake into a melter, which heats the plastic to 200°F (93°C), and then adds dyes to color the new material. Another machine extrudes the melted material to form small pellets about a quarter-inch (0.64 cm) in diameter, and the pellets then cool. (HDPE can also be formed into powders, granules, tubes, or sheets.) Manufacturing plants then use the HDPE to make new plastic products. Recycled HDPE does not disintegrate in heat or humidity, splinter, or lose its color, so manufacturers have preferred it for making bottle caps, outdoor furniture, playground equipment, toys, crates, dog houses, and boating parts.

The United States now recycles more than 27 percent of HDPE bottles. By 2005, the total amount of plastics going into recycling had for the first time exceeded 2 billion pounds (907 million kg) per year and the rate of recycling—the amount recycled divided by the total amount of plastic—continues to grow. The German market research company Ceresana Research reported in 2008 that the growing global HDPE market had exceeded 30 million tons (27 million metric tons), and HDPE revenues would likely double by the year 2016.

The most important aspect of HDPE recycling resides in its capacity to conserve petroleum. It takes 3.86 pounds (1.75 kg) of petroleum to make 2.2 pounds (1 kg) of new HDPE when accounting for the raw materials and the energy to run the process and transport the goods. Plastic recycling also requires some energy for the recycling process, transportation, and manufacturing, but if these tasks are done efficiently, plastic recycling can save energy compared with making new plastic.

ers chemicals that dissolve in either oily liquids or water. Catalyzed reactions consist of chemical reactions in which a compound called a catalyst aids the progress of the reaction. The resulting reaction products differ from the starting compounds.

Recycling chemistry has advanced from the time of the first environmental gatherings in the 1970s. At that time recycling had not become the familiar activity it is today in almost every home, school, and business. Though recycling of some sort has existed for centuries, only in modern history have advanced chemical processes been used to break apart materials and devise new substances. The sidebar "Case Study: Recycling during World War II" on page 46 examines one of the early successes in recycling in U.S. history.

MINERALS AND METALS

Minerals and metals are two nonrenewable resources that must be recycled to sustain the current amounts of these substances in the Earth's crust. Minerals consist of compounds with specific crystal structures and other physical features that identify them, such as density, hardness, color, luster, and ability to break. Mineralogists have to date identified about 4,500 minerals in the Earth's crust. The following are examples of the most plentiful minerals: quartz, feldspar, mica, olivine, calcite, and magnetite.

Mineral recycling plays an important role in protecting the environment in four main ways. First, it avoids mining activities that destroy land inhabited by endangered plants and animals. Secondly, it reduces the amount of toxic wastes produced by mineral mining. Third, the process of extracting a mineral from ore, the raw material removed from a mine,

METALS		
GROUP	**DESCRIPTION**	**EXAMPLES**
base metals	corrodes when exposed to air and reacts with hydrochloric acid to release hydrogen gas	copper, iron, nickel, lead, zinc
ferrous metals	usually magnetic	iron
noble metals	resists corrosion when exposed to air	gold, platinum, rhodium, silver
precious metals	rare and of high monetary value	gold, silver, palladium, platinum, plutonium, uranium

requires additional hazardous chemicals that recycling does not require. Lastly, mining and mineral extraction consumes 10 percent of the world's energy, which is disproportionately large for the size of this industry. Mineral recycling alleviates this consumption.

A metal consists of an element in which electrons move easily between the atoms and help bond the atoms together. Metals have characteristic density, luster, and conductivity of heat and electrical charge. Metallurgists classify elements in various ways based on chemical qualities. Members of the periodic table of elements often share characteristics, so chemistry classifies these elements into more than one category. For example carbon in the form of graphite conducts electrical current like a metal, so shares this characteristic even though carbon is not a metal. The table on page 44 summarizes four main metal classifications.

Chemists also classify metals by their electron configurations. The periodic table of elements reflects these groupings as follows:

- Group IA: lithium, sodium, potassium, rubidium, cesium, francium
- Group IIA: beryllium, magnesium, calcium, strontium, barium, radium
- Group IB: copper, silver, gold
- Group IIB: zinc, cadmium, mercury
- Transition metals: scandium, titanium, vanadium, chromium, manganese, iron, cobalt, nickel, copper, zinc, yttrium, zirconium, niobium, molybdenum, technetium, ruthenium, rhodium, palladium, silver, cadmium, hafnium, tantalum, tungsten, rhenium, osmium, iridium, platinum, gold, mercury

Metal recycling methods differ slightly by the individual metal to be recovered, but most recycling includes the following basic steps: washing, shredding or chopping, purification, melting, and casting or molding into blocks called ingots. The metals industry has traditionally used smelting as a purification process. Smelting involves heating an ore to melting so that impurities can be separated from the desired metal.

Metal recycling has grown into a critically important part of the metals industry and the U.S. economy. For instance, the metals industry

CASE STUDY: RECYCLING DURING WORLD WAR II

World War II began in September 1939 when Germany invaded Poland. From that point on, the war enveloped an increasingly large area of the globe (the United States entered in 1941). Although the United States held a vast potential in workers and natural resources, it still needed some raw materials from places halfway around the world. As industries converted their normal activities into wartime efforts, certain materials ran short for meeting both civilian and military needs. The United States and many other countries turned to massive recycling programs to generate the most needed and scarcest items.

During World War II in the United States, people everywhere, from big cities to small towns, began collecting rubber tires, cloth, motor oil, and various metals in what came to be known as scrap drives. These scrap drives and a complementary program in food rationing helped conserve items that were no longer available from overseas because of Japanese and German naval operations in the Pacific and Atlantic Oceans, respectively. A rationing campaign set household limits on things such as sugar, coffee, meat, eggs, fish, cheese, shoes, and gasoline. Eggs, milk, and meat came from the United States, but the fuel needed to transport them had been redirected into the war effort.

U.S. recycling campaigns in World Wars I and II netted thousands of tons of materials for supplying the war. The scrap drives of World War II, pictured here, led to the development of new chemistry and materials science methods in use today. *(Bentley Historical Library, University of Michigan)*

As the war progressed, the scrap drives expanded. Old rubber tires initially helped supply some of the material for military vehicle tires, but soon President Franklin D. Roosevelt called for "old rubber raincoats, old garden hose, rubber shoes, bathing caps, gloves" to supplement the drive. The historian Ronald H. Bailey described the colorful outcomes of some scrap drives in *The Home Front*, "In New York City, a carload of chorus girls from a Broadway musical drove up to one collection depot,

Nat Jupiter's service station, and wriggled out of their girdles. In Washington, Secretary of the Interior Harold L. Ickes spotted a rubber floor mat at the White House, rolled it up and had his chauffeur deposit it at the nearest collection point." Towns received quotas from the government setting the amount of material the towns were required to put into the collection trucks, and the populace readily complied.

The recycling campaign of World War II showed that with ingenuity almost any scarce article seemed to have a wartime purpose. Cloth scraps of almost every type such as used clothes, rags, curtains, blankets, and old upholstery went into uniforms. Women's silk stockings disappeared from store shelves because the silk and newly invented nylon went to making parachutes. World War II changed currency forever by rerouting all copper from penny production to wire production. Pennies for a time contained zinc-coated steel, but no copper. Used lubricating grease went into the manufacture of explosives and artificial rubber; food grease and fats helped in making gunpowder. Old newspaper served as packaging for shipments to the military.

Of all the war's recycling drives, those for scrap metals, especially iron, and rubber made the biggest impact on wartime machinery. Huge heaps of metal items grew in the center of almost every town and contained bicycle rims, watering cans, metal drums, cans, paper clips, toy wagons, pipes, bedsprings, among other household products. Haulers took the metals to smelting companies that removed the impurities by heating the metal until it was molten. Trucks then carried the cooled, extracted metal to factories.

Equally impressive collections of rubber objects mounted up in World War II: rubber soles, rubber bands, balls, roofing liners, inner tubes, and other items that President Roosevelt asked for in his weekly radio addresses to the nation. Rubber recycling had not previously been done on such a large scale, but wartime chemists soon found ways to improve the process for turning rubber into new products. They used a process called vulcanization in which rubber is heated and reacted with a sulfur compound. Vulcanization degrades and then rebuilds the links between the rubber's molecules to strengthen the rubber. The rubber industry also uses *devulcanization* in rubber reprocessing in which the sulfur is removed in order to change its chemical properties. A product manufacturer then revulcanizes the rubber, which returns durability to it.

During World War II, the United States recycled about 25 percent of its total wastes, an enormous amount for a country of about 138 million people. The scrap drives helped industry develop new, faster ways to recycle raw materials and led to a variety of innovative uses for commonplace items.

accounted for more than 25 percent of the gross output in dollars of all U.S. durable goods industries in 2009, and recycling made up a significant part of this production.

RUBBER RECYCLING

The United States discards 250 million tires annually, a portion of which can be recycled to make new tires or other rubber-containing products. Without recycling, tires create an unmanageable volume of nondegradable waste. Mountains of discarded tires have recently been shown to cause health hazards by filling with rainwater, which provides breeding grounds for insects that carry diseases such as West Nile virus and encephalitis. Also, fires at tire dumps have released large amounts of air pollution containing hazardous chemicals.

Tire and other rubber products recycling begins with the pulverization of the rubber into fine granules called crumbs. Crumb rubber represents the primary raw material from recycling that goes to product manufacturers. The next step is devulcanization, whereby a chemical process breaks the sulfur linkages that hold rubber's polymers together so that it can be remolded into a new product. Devulcanization has been difficult for recycling companies to perfect; the current methods either are very expensive or ruin some of rubber's natural qualities.

The rubber recycling industry has experimented with new technologies that carry out devulcanization in ways other than relying on a difficult-to-manage chemical process. Ultrasound devulcanization, for example, exposes the rubber to ultrasonic waves of 50 kilohertz (kHz) for

The latest scrap tire estimate from the EPA in 2003 calculated that Americans generated 290 million scrap tires that amounted to 2 percent of the entire U.S. solid wastes. About 30 states use tire incinerators for the production of part of their energy needs. The energy is called tire-derived fuel (TDF). *(Buffalo ByProducts)*

20 minutes. This treatment breaks sulfur bonds while leaving all of the rubber's carbon-carbon bonds intact, so the resulting rubber maintains its desired qualities.

The world demand for rubber has been increasing, mainly because of dramatic increases in vehicle use in China and India. Current recycling methods contribute to only a small fraction of this growing demand. For this reason, the recycling industry as well as tire manufacturers continue to seek new technologies in rubber treatment and reuse.

CONCLUSION

Waste recycling has historically been a means to either hold down business costs or to invent a replacement for something that has become scarce, or both. Recycling has clearly taken on a third purpose in the past several decades: waste management. The volume of waste that the world's population produces is an environmental problem that recycling helps reduce. In order for recycling to make a greater impact on waste reduction, there are some areas of recycling technology in need of attention.

Since 1970, the recycling industry has made advances in the number of materials it reuses and the variety of products made from these materials. Materials science and chemistry will in the future create additional uses for the various items that continue to accumulate in landfills. In general, landfill items are composed of glass, plastic, aluminum, nonaluminum metals, paper, and cardboard. Other landfill items either decompose quickly or can be removed and made into fuel for energy production. Science must find more innovations for the five main recyclable materials produced today by communities. Research also needs to find solutions for dealing with chemicals and solvents in order to make the best use of these sometimes hazardous substances.

Recycling's greatest accomplishment may be in aluminum recycling, an efficiency that hardly leaves room for improvement. Other materials have not been converted to usable products as efficiently, perhaps because the technology is lacking. Plastic recycling in particular has not advanced very well. The plastics industry often finds that producing plastic resins from new raw materials is less costly than recycling certain resins. Plastics present an obvious opportunity in recycling technology.

Finally, though recycling now serves a purpose in managing wastes, the future will demand improved technologies to keep up with waste accumulation. The recycling industry might prove to be the best arena for developing a zero waste society. Zero waste—though a long-term goal—has the potential of reducing humanity's ecological footprint, which is the ultimate goal of all sustainability programs.

GASOLINE ALTERNATIVE VEHICLES

In the United States, drivers log 4 million miles (6.4 million km) annually on highways and streets in personal vehicles, a distance exceeded only by air travel. Worldwide passenger vehicle production reached almost 52 million in 2010; the number of vehicles produced worldwide has been climbing since the 1950s. Alternative travel advocates feel that the only way to reduce global warming is to forsake this massive dependence on personal vehicles. But environmental scientists might do better to accept the reality of the strong personal connection between people in industrialized nations and their cars. Industrialized nations depend on commuters and also on trucks that move products from manufacturers to customers.

Mike Millikin of the Green Car Congress and environmental writer Alex Steffen wrote in *Worldchanging: A User's Guide for the 21st Century* in 2006, "For many North Americans, the car has become both a necessity and a shrine. Sprawling suburbs and bad urban planning have made it nearly impossible for us to get anywhere without driving." Booming economies in China and India among other regions have followed the United States in their desire for new cars. The *Washington Post* reporter Ariana Eunjung Cha noted in 2008, "Car ownership in China is exploding, and it's not only cars but also sport-utility vehicles, pick-up trucks and other gas-guzzling rides . . . China alone accounts for about 40 percent of the world's recent increase in demand for oil, burning through twice as much now as it did a decade ago." Car manufacturers have increased launches of new models in India to meet that country's growing demand for personal vehicles, and the trend does not appear to be slowing.

Environmental scientists have taken their cue from these statistics to assume that getting people to give up their cars will be extremely difficult—in some cases impossible. But the long-term future of fuel for these vehicles presents another worry. World oil consumption continues to increase, led by the United States, which consumes more than 20 million barrels daily, followed by China, Japan, Russia, Germany, India, Canada, Brazil, South Korea, Saudi Arabia, Mexico, France, United Kingdom, Italy, Iran, Spain, and Indonesia, which all consume more than 1 million barrels a day. Rather than trying to alter people's desire for fuel consumption, energy technologies must develop new fuels to replace nonrenewable petroleum.

Fossil fuel consumption inevitably leads to greenhouse gas emissions from exhaust. Transportation produces 34 percent of the total greenhouse gases—electric power plants produce 39 percent and homes and industries produce 27 percent. For this reason, cleaner fuels and more efficient fuel use in vehicles can have an important effect on global warming caused by greenhouses gases. On roads today, the main producers of greenhouse gases are cars (35 percent of emissions), light trucks (27 percent), and heavy trucks (19 percent). (Aircraft produce 9 percent of greenhouse gas emissions, and pipelines, locomotives, and ships and boats produce the rest.)

The main alternative fuels today are ethanol, biodiesel (nonpetroleum diesel fuels), natural gas, propane, and hydrogen. Vehicles running on either ethanol or biodiesel currently make up the largest percentage of alternative fuel vehicles, and a small selection of models have been introduced that run on natural gas, propane, or hydrogen, with additional models soon to be introduced. Even so, alternative fuel vehicles make up a very small proportion of the total vehicles bought in the United States, about 2 percent of new car sales.

This chapter discusses the important technology of alternative fuel vehicles, from electric-gas *hybrid vehicles* that have already entered the market to innovative vehicles still in development. The chapter examines new ideas in *biofuels,* synthetic fuels, and other power sources such as batteries, fuel cells, and natural gas. Sections discuss the vision for a future technology and an ancient technology, nuclear-powered transport and wind power, respectively. The chapter closes with a discussion on the feasibility of new vehicles, based on hybrid technology that has already been successful.

Car designers, engineers, and amateur inventors have developed a number of truly innovative alternative fuel vehicles. This solar-powered car designed and built by Kansas State University consists of an upper body covered with solar collectors and batteries. More than 40 universities in North America have designed similar cars as prototypes for future all-solar or solar-electric vehicles. The teams also compete in a biennial 2,500-mile intercollegiate North American Solar Challenge, a cross-country rally.

EVOLUTION OF ALTERNATIVE VEHICLES

Today's alternative vehicles include any vehicle with a power source that either replaces gasoline or conserves gasoline by sharing the power needs with another type of energy, such as electricity. The introduction of the Toyota Prius in 1997 marked the first mass-marketed hybrid car for family use, but the vision for alternative vehicles dates much further into the past. The following table reviews important milestones in the history of modern alternative vehicles.

In the 1930s, the electric vehicles that had dominated the early 20th century were replaced by gasoline vehicles. Electricity was a technology that offered limitless uses in the home, but for vehicles it presented the following troubles: Electric power plants were not standardized to using either AC or DC voltage; the range between battery recharges lasted only 30–50 miles (48–80 km); batteries lost about 40 percent of their power in the winter; and heavy batteries made vehicles get stuck in snow and mud.

The development of alternative vehicles has been a steady series of trials and errors. Each promising breakthrough in a new fuel to replace gasoline has been accompanied by unique drawbacks. In 2008, the *Time* magazine reporter Michael Grunwald warned readers of the pitfalls of putting hopes on biofuels as the perfect answer to fuel consumption: ". . . the biofuel boom is doing exactly the opposite of what its proponents

EVOLUTION OF ALTERNATIVE VEHICLES

VEHICLE	INVENTOR	DATE	FEATURES
diesel engine	Rudolf Diesel	1890s	first engine to run on peanut oil, a precursor to today's biofuel
electric carriages		1898–1912	all-electric powered vehicles predominated
Porsche-Lohner	Ferdinand Porsche	1900	electric drive motor worked with a gasoline-powered engine
Model T	Henry Ford	1905	originally designed to run on ethanol from corn
Cadillac	Charles Kettering	1912	first electric starter on vehicle
small-scale car	Christopher Becker	1935	all-electric car
various vehicles	Ford Motor Company	1930s–40s	offers alternative fuel Benzol
road vehicles	Thomas Davenport and Robert Davidson	1942	non-rechargeable electric cells for road use
Electrovan	General Motors	1966	hydrogen fuel-cell powered
trucks	Ford Motor Company	1960s	some models powered by propane
CitiCar	Sebring-Vanguard Company and Elcar Corporation	1970s	small, short-range electric cars
Prius	Toyota	1997	first successfully marketed electric-gasoline hybrid

Electric-gasoline hybrid and all-electric vehicles will become more popular in the near future. Many city and university car-sharing programs use vehicles similar to this Subaru R1e plugged in to a charger, and several car companies now offer all-electric models. *(Subaru)*

intended: it's dramatically accelerating global warming, imperiling the planet in the name of saving it." Biofuels, which a mere decade ago seemed to signal the future of alternative fuels, now receive the most criticism due to the negatives that come with its positives. The following table highlights the advantages and the disadvantages of alternative fuels as scientists and engineers explore these options in greater detail.

Automotive engineers work on ways to improve the current advantages of alternative fuels while trying to eliminate the disadvantages. Most of the innovations described in the table have been put into *prototype* vehicles, some of which might enter the consumer market in the near future. One of the biggest obstacles in implementing new vehicle designs comes from breaking old traditions in automaking and buying. Large automakers have in the past built their financial growth on gasoline-fueled vehicles and have put relatively little effort into alternative vehicles. When crude oil prices were low and air pollution had not yet reached critical levels, gasoline made sense. But the air now reflects the damaging effects of tons of vehicle emissions, and crude oil supply has turned into a complex scientific and political problem. The decision not

CHARACTERISTICS OF ALTERNATIVE ENERGY SOURCES FOR VEHICLES

FUEL	ADVANTAGE	DISADVANTAGE
battery	nonpolluting	limited range at present
biofuel (corn ethanol)	can use gasoline pumps	disrupts crop prices and world food supply
biofuel (non-corn sources)	large supply, works in diesel engines	potentially high carbon dioxide emissions
electricity, plug-in	nonpolluting	depends on availability of remote plug-in sources
hydrogen fuel cell	produced from water, nonhazardous, and no carbon dioxide emissions	requires energy to produce it; short driving range at present
natural gas	high energy yield; low cost	nonrenewable
solar	nonpolluting	expensive; not feasible for near future
synthetic fuel	large supply	high manufacturing cost and environmental impact

to develop new types of vehicles has caused dire consequences for the U.S. automobile industry, which is discussed in the sidebar "Case Study: Toyota's Prius" on page 58.

Truck manufacturers have also made headway in converting today's fleet of long- and short-haulers to fuel-efficient vehicles. Flex-fuel trucks, for instance, run on gasoline, ethanol, or hydrogen, similar to the options in hybrid cars. The trucking industry also has followed guidance from the U.S. Environmental Protection Agency (EPA) in retrofitting the current fleet of diesel-burning trucks, buses, and construction vehicles with anti-pollution technology. Some of the technologies that may help lower truck emissions include:

- engine idle reduction to conserve fuel
- improved catalytic converters to reduce harmful emissions
- catalyst mufflers to clean the exhaust
- particulate filters to remove particles from the exhaust

The trucking industry also encourages its drivers to reduce long periods of idling and manage their speeds to cut the total emissions they produce.

BIOFUELS

Biofuels are any fuels that are made from plant material. The main biofuels in use today are ethanol produced from grain crops; methanol produced from natural gas or from solid organic waste called biomass; biogas, a mixture of methane and carbon dioxide (CO_2); and vegetable oils. Mounting anxieties over global warming due to vehicle emissions and the precarious supply of crude oil affected by political differences have made biofuel a priority in the United States. George W. Bush emphasized the need for biofuels in the 2007 State of the Union address: "Let us build on the work we've done and reduce gasoline usage in the United States by 20 percent in the next 10 years. . . . To reach this goal, we must increase the supply of alternative fuels, by setting a mandatory fuels standard to require 35 billion gallons [132 billion l] of renewable and alternative fuels in 2017." Both political parties now are concentrating on the biofuel mandate. Biofuel producers have taken up the challenge to increase operations and worldwide investment in biofuels has grown. Biofuel investments may top $100 billion by 2010. However, even as production has soared, so too has concern over the millions of acres being converted from food crops to biofuel crops.

Ethanol has occupied the center of the controversy of converting food crops to fuel crops and its effect on world economies. Biofuel can be made from corn, soybeans, sugarcane, sugar beets, sorghum, or sunflowers. Increased prices for these crops as *feedstock* for ethanol production have the potential of inducing farmers to sell crops to fuel producers rather than to food producers. Additional growers see the good prices they can get from growing crops such as corn for biofuel, so they convert their crops to corn also, causing other grains to rise in price. A global demand for crops ensues, and subsistence farmers in developing countries clear forests and grasslands to plant crops. As a result, habitat and biodiversity disappear, and the cutting and burning of land for cultivation adds CO_2 to the atmosphere. This series of events has played out in a few places already, such

CASE STUDY: TOYOTA'S PRIUS

When Toyota's electric-gasoline hybrid vehicle went on sale in the United States in the summer of 2000, the *New York Times* reporter Andrew Pollack wrote, "The Prius, a so-called hybrid that uses both gasoline and electric power, avoids most of the drawbacks and inconveniences of other vehicles that are designed to be clean and fuel-efficient." Pollack pointed out an often-overlooked fact: The Prius was not the first hybrid vehicle to arrive on the automotive scene. The Japanese automaker Toyota built on previous experiments in electric-gasoline vehicles that dated as far back as 1900. But Toyota achieved success by making the Prius the first marketable hybrid vehicle that met the needs of drivers while staying ahead of increasingly strict environmental laws.

The Prius symbolized a feature of Japan's automotive industry that began in the 1980s and continues to this day. That is, Japanese cars—as well as Japan's dominant electronics industry—combined driver-friendly innovations with good economic decisions for controlling manufacturing costs and pricing cars for the average car buyer. American automakers have developed as many if not more alternative vehicle prototypes than their foreign competitors, but they have not achieved all of the five following objectives that the Prius achieved:

- a car with conventional look and feel that did not require drivers to greatly change their driving habits
- comparably priced with similar cars
- saves consumers money to operate
- significantly better for the environment than the competitors being sold at the time
- readily available on the consumer market

as Brazil, and environmentalists fear it may reach critical proportions. The *Time* magazine writer Henry Grunwald explained simply, "The basic problem is that the Amazon [in Brazil] is worth more deforested than it is intact." Biofuels should not be canceled as alternative fuels. Rather, biofuels must be managed to work better with the environment.

New biofuel sources may help ease the problems associated with corn-dominated biofuels. Growers have already begun to experiment

Despite innovations by the Big Three U.S. automakers—Ford, General Motors (GM), and Chrysler—that preceded the Prius, these companies have been playing catch-up in the alternative vehicle market ever since the Prius introduction. In the Prius's early years, the Detroit automakers seemed determined to follow their tried-and-true approach of building bigger, more powerful vehicles that appealed to many drivers but also guzzled gasoline.

By 2008, the Big Three were experiencing serious financial pressures, partly due to an obsolete product line. A *New York Times* article explained the automakers' predicament in reinventing their business with a goal of sustainability: ". . . the car companies, which have long lead times to develop products, will need sales of big trucks and sport utility vehicles . . . to bring in much-needed revenue." This statement highlights the challenges ahead for the Big Three in attaining three sometimes conflicting objectives. First, they must shorten the time between designing a new vehicle and selling it. Second, they must design innovative and desirable alternative vehicles. Third and most daunting, the Big Three must take the risk of changing their current offering of vehicles and trust car buyers will accept them.

The lessons to be learned from the Prius have extended beyond fuel economy. Progress in sustainability does not require a drastic overhaul of current technology, but it must lead to a few important changes that make a genuine difference to the environment: type of fuel, fuel efficiency, and styling to minimize fuel use. These innovations should also appeal to the public. A prototype vehicle that makes brilliant use of energy but has no practical application for families is all but useless in building sustainability. Finally, automobile manufacturing must restructure in order to take inventions from the planning stage to consumers much faster than in the past. Given a choice between large, gas-guzzling vehicles and more efficient and affordable alternatives, many drivers will choose the latter. For this to happen, carmakers must be committed to alternative vehicles. They must not treat these vehicles as novelties but view alternative vehicles as the future of transportation.

Sugarcane, shown here, forms the basis of Brazil's biofuel program, allowing Brazil to no longer rely on foreign oil. Sugarcane produces twice the biofuel per acre as corn, the main feedstock for U.S. biofuel. All biofuel and biomass feedstocks must be grown sustainably and preserve the world's food supply and the environment. Brazil burns sugarcane fields—and pollutes the air—to drive out snakes before workers enter the fields to be harvested. The cutting of Brazilian forests to plant more sugarcane puts additional CO_2 into the atmosphere. The biofuel industry must solve these problems in order to truly help the environment. *(Rufino Uribe)*

with crops that make better use of land and convert crop energy to fuel energy more efficiently than corn. While corn generates 1.3 units of ethanol energy for every 1 unit of corn processed, sugarcane yields 8 units of ethanol per 1 unit of sugarcane. Sugarcane yields can also double corn yields per acre [0.004 km²] of land. Any crop used for biofuel production should be chosen to avoid the heavy use of chemical fertilizers, pesticides, and herbicides that have become a hallmark of large agriculture's corn production.

Waste materials also offer efficient conversions of the energy stored in waste to energy contained in ethanol. Cellulosic ethanol comes from substances that have little crop value; they are called cellulosic because they contain mainly cellulose fiber. Cornstalks, husks, leaves, forestry wastes such as wood chips and bark, sawdust produced by lumber mills, paper pulp, and fast-growing prairie grasses such as switchgrass produce 36 units of ethanol energy per unit of material. Growers in Mali in eastern Africa where good agricultural land is at a premium have begun growing jatropha, a plant that thrives on poor soils and requires little fertilizer and no pesticides to produce high yields. This type of crop allows farmers to keep more valuable cropland for producing fruits and vegetables and income.

Some car owners have taken matters into their own hands by retooling their cars to run on waste cooking oil. Greasecar is a Massachusetts company that sells kits for car owners to use in modifying an engine for running on vegetable oil left over after use by restaurants. At present, however, the EPA has not approved these modifications or vegetable oil fuels for use on public roads.

Entrepreneurs have also studied *algae* as a renewable energy source that requires no cropland at all. GreenFuel Technologies in Cambridge, Massachusetts, grows algae in ponds, taking advantage of algae's ability to turn Sun energy into carbohydrates and fats using photosynthesis. Company chemists then convert 100 percent of the substances into ethanol. *Fortune* magazine reported on GreenFuel's system in a 2008 article: "The curious setup is a bioreactor [a microbe-growing tank] that takes the stuff of pond scum—algae—grows it like mad, and turns it into biomass that can be processed into fuel for cars and trucks." Algae tanks can furthermore be built on land that is poor for cultivation, and algae even grow on polluted or salty waters.

The small community of scientists that have worked on techniques for making fuel from algae have targeted algae biodiesel. Diesel fuel is derived from crude oil, as is gasoline, but diesel differs from gasoline in composition and has a thicker, oilier consistency. Biodiesel comes from plant sources rather than crude oil and produces 2.5 times the energy produced by an equal amount of fossil fuel. Though the research remains in its early stages, algae seem best suited for making biodiesel.

Algae and other microbes, such as bacteria, have created excitement among scientists as a new alternative fuel source that bypasses

Algae convert solar energy to chemical energy, which companies use as feedstock for biodiesel production. Algae are reliable, inexpensive to grow and harvest, and need a fraction of the space that agricultural crops use. *(New Mexico State University)*

the disadvantages of growing fuel crops. The biofuels researcher Kathe Andrews-Cramer of Sandia National Laboratory in New Mexico has said, "Algae have the potential to produce a huge amount of oil. We could replace certainly all of our diesel fuel with algal-derived oils, and possibly replace a lot more than that." Power companies and other large businesses are now conducting studies on algae-produced biofuels for the future. Large oil companies, some of the biggest companies in the world, have partnered with entrepreneurs in seeking new biofuels.

SYNTHETIC FUELS

Synthetic fuels, also called synfuels, consist of liquid fuels derived from nonliving things. The main synfuel sources currently being studied are coal, natural gas, oil shale, and tar. Synthetic fuels came to prominence in World War II when crude oil supply lines were cut. Germany developed the method known as the Fischer-Tropsch process for making synfuel as its fuel supply dwindled. The Fischer-Tropsch process makes liquid gasoline from coal by applying high temperature and high pressure to the solid coal. The first synfuels came from coke as the starting material; coke is a by-product made by distilling coal. By the 1950s, crude oil again became available and research on synfuels slowed. Alternative vehicle technology has reignited interest in synfuels for two reasons: to break U.S. dependency on crude oil supplied by other countries and to give vehicles a cleaner-burning fuel with reduced exhaust emissions. Synfuels have one major drawback, however, because their manufacture requires high levels of consumption of fossil fuels and energy. For this reason, synfuel producers have explored the use of biomass to substitute for fossil fuels as raw material.

Biomass works as a feedstock in synfuel production because it contains a high concentration of carbon compounds, the basis for fuels used in combustion. The synthesis reactions produce long hydrocarbons made up of hydrogen molecules attached to a carbon backbone. Because biomass composition is variable, it results in many different fuels, each with a unique blend of hydrocarbons of varying length. The Fischer-Tropsch process applied to biomass makes the following materials, from the least dense to the densest: methane gas, ethane gas, liquid petroleum gas, gasoline, diesel, and waxes.

Many motor oils have synthetic versions that drivers use today. Synthetic motor oil contains long polymer compounds made in laboratories, chemically designed to behave as regular oil. The synthetic oils must not degrade when heated in an engine and should offer lubricating qualities equal to or better than regular motor oil. So far, chemists have developed synthetic oils in a variety of grades or viscosities that meet the needs of different types of engines.

Biotechnology companies have also joined the hunt for better sustainable synfuels by combining biology with the synthesis process. A new field called synthetic biology involves the fabrication of biological substances not found in nature. Kareem Saad of Codon Devices explained in 2008, "Synthetic biology is important for a lot of reasons. Introducing engineering principles of design . . . and standardization to biology promises to revolutionize the way we make fuels and consumer products that rely less on crude oil and are less damaging to the environment, and can be a game-changer." Synthetic biology experts now plan on manipulating, through *bioengineering* or through new *fermentation* methods, natural microbes to make new hydrocarbons that act as fuel. A new generation of biofuels might be synthesized by microbes to produce new hydrocarbons to make synthetic gasoline or synthetic diesel.

Green chemistry represents a related field in which chemists use biological components such as enzymes to carry out chemical reactions. Natural enzymes run reactions without demanding high temperatures or hazardous chemicals that conventional chemistry sometimes uses. This field of chemistry offers the similar promise that synthetic biology does for creating hydrocarbons that take the place of fossil fuels.

Whether a vehicle uses a biofuel, synfuel, or fossil fuel, a combustion engine must have hydrocarbons to burn for producing power. Alternative fuel combustion engine vehicles therefore resemble standard gasoline-powered vehicles. The principles of combustion are described in the following sidebar "Combustion."

BATTERY POWER

Vehicles powered exclusively by batteries have been tried for many years. Most progress toward this end has been stymied, however, by the enormous size and weight of batteries that would be needed to power cars for any practical distance. The development of gasoline-battery hybrid vehicles has offered more promise and has spurred scientists to improve battery technology for vehicles. With newer, lighter batteries, vehicles completely powered by batteries may become a large part of the alternative vehicle market.

Conventional car batteries contain lead and strong acid that provide a medium for the flow of electrons between two oppositely charged poles. This electron flow becomes electrical current, which helps start the engine when a driver turns the key. In the 1990s, GM developed a new type of battery not based on the usual lead-acid system. The company used their invention to create the all-battery-powered EV-1 car. The EV-1 performed as a conventional car with the same or better speed and power, but recharging the battery proved to be impractical at the time, and the company discontinued the EV-1 in 1999.

Two innovations aided the return of battery-powered vehicles. The first belonged to Toyota's Prius with its new lighter battery. The second innovation came from the U.S. computer industry's development of

COMBUSTION

Combustion is a process in which oxygen combines with other atoms to make a new compound and gives off energy as heat. Fire is a well-known combustion reaction. In fire, oxygen from the air combines with carbon or carbon-hydrogen compounds and produces heat, gases, and incandescence, the light emitted from a heated material.

The internal combustion engine makes this process possible. Inside an internal combustion engine's strong metal chambers, called cylinders, rapid reactions occur between the high-energy fuel (gasoline) and air. These reactions are actually energy-releasing explosions when hydrocarbons and air combine. The rapid series of explosions provides the power to move a vehicle forward by turning a crankshaft. In short, gasoline-powered vehicles move because the energy held in gasoline's carbon-hydrogen bonds has been converted to heat, and the heat energy is in turn converted to kinetic energy.

Typical internal combustion engines make conversions of one type of energy to another in four steps: (1) the intake stroke in which gasoline enters the combustion chamber with air; (2) the compression stroke that puts pressure on the gasoline-air mixture to make the explosion more powerful; (3) the combustion reaction in which the explosion occurs within the cylinder; and (4) the exhaust stroke in which the reaction's by-products exit the cylinder. Biofuels and synfuels contain hydrocarbons that work just as well as gasoline hydrocarbons in the combustion reaction. Furthermore, exhaust by-products from biofuels and synfuels do not carry the same high concentration of hazardous substances as gasoline exhaust.

The exhaust from combustion engines is blamed as one of the major contributors to global warming. Gasoline exhaust contains compounds that threaten human, wildlife, plant, and tree health by causing global warming: carbon monoxide, nitrogen dioxide, sulfur dioxide, and polycyclic hydrocarbons (hydrocarbon compounds containing ring structures also made of carbon and hydrogen). The exhaust from conventional combustion engines also contains benzene, formaldehyde, and small particles less than 10 micrometers in diameter. Biofuels and synfuels therefore preserve human and environmental health by reducing or eliminating the production of hazardous exhaust components.

lithium-ion batteries for portable computers. Both of these types of bat-teries generated an acceptable amount of power, and they weighed far less than the batteries of the past. All subsequent alternative fuel vehicles now contain components, including the batteries, that have been selected both for durability and for weight.

The newest generation of battery-powered vehicles are hybrids similar to the Prius or fully rechargeable models that are plugged in when not in use. The *Chicago Tribune* reported in 2008, "General Motors Corp. and Toyota have announced plans to introduce plug-in hybrids in 2010, and both will use lithium-ion batteries." Describing GM's new Volt, the article said, "After the batteries drain, the Volt's gas engine recharges them, add-ing another 600 or so miles [965 km] to the vehicle's range." Automakers will need to continue improving the range of all-battery-powered cars so that drivers never fear they will be stranded by a dead battery with no recharging outlet near. Engineers who plan new sustainable systems in cities have likely considered the need for downtown plug-in terminals for recharging cars during the workday.

An auto industry spokesman John Hanson told the *Chicago Tribune,* "We need to see how lithium-ion batteries perform in the real world and make sure this technology is robust and what they [car companies] need." Automakers have reached a point where they can go in several directions: toward cars powered completely by lithium-ion batteries; by new battery technology that works better than lithium-ion; or by a combination of gasoline and battery in hybrid vehicles.

FUEL CELL TECHNOLOGY

Fuel cells represent a new phase in battery power. Fuel cells produce only water and heat in their energy-generating process, they are quiet, and they convert fuel to energy three to four times more efficiently than combustion.

The first of two types of fuel cell reacts hydrogen gas (H_2) with oxy-gen gas (O_2) to produce electrical energy. Automakers expect to introduce vehicles dependent on this hydrogen fuel cell technology between 2010 and 2020. The second type of fuel cell uses biological reactions supplied by microbes that carry out the same reaction between hydrogen and oxygen to create a flow of electrons. Scientists have been developing biological fuel cells, but so far they have not been tried in vehicles.

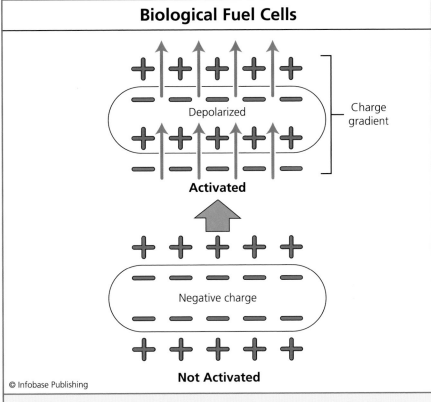

Biological Fuel Cells

Charge gradient

Activated

Not Activated

© Infobase Publishing

Mechanical fuel cells can be based on the activities of biological fuel cells, such as the cell shown in the diagram. Biological cells generate energy as an electric current by creating a charge gradient. A resting cell is depolarized: The cell membrane separates the negatively charged interior (proteins) from the positively charged exterior (sodium ions, Na+). By opening the membrane's pores, the cell depolarizes and a charge gradient develops, which sets up a current.

Both chemical and biological fuel cells rely on a catalyst, which is any substance that makes helps reactions proceed by lowering the energy needed to start the reaction. The catalyst enables the following hydrogen fuel cell reaction to take place in an efficient manner:

$$H_2 \text{ fuel} + O_2 + \text{catalyst} \rightarrow H_2O + \text{energy}$$

Chemical fuel cells use metal catalysts that readily give up or accept electrons, such as palladium and platinum. Biological fuel cells do the same using enzymes, which act as catalysts in reactions that occur in nature. In biology, catalysts enable reactions to take place in milliseconds, or thousandths of a second. Without enzymes, the same biological reactions could well take several million years. In 2003, the biochemistry professor Richard Wolfenden of the University of North Carolina explained, "Now we've found one [a reaction without enzymes] that's 10,000 times slower than that. Its half-time—the time it takes for half the substance to be consumed—is 1 trillion years, 100 times longer than the lifetime of the universe. Enzymes can make this reaction happen in 10 milliseconds." Clearly, the success of fuel cell technology depends on catalysts, and biology may have already invented some of the best catalysts on the planet.

Fuels for chemical fuel cells may be any of the following hydrogen-rich materials: natural gas, petroleum, propane, methanol, ethanol, or coal. Only methanol and ethanol from this list are renewable energy sources. Even though chemical fuel cells use nonrenewable fuels, they convert fuel to energy much more efficiently than combustion engines. Chemical fuel cells also reduce the amount of CO_2 emissions by about two-thirds.

Biological fuel cells, by contrast, can use biomass or organic wastes, such as manure, as a fuel source. Although Wolfenden pointed out the

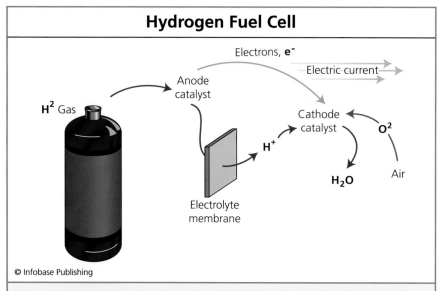

Hydrogen Fuel Cell

Electrons, e^-

Electric current

Anode catalyst

H^2 Gas

Cathode catalyst

O^2

H^+

H_2O Air

Electrolyte membrane

© Infobase Publishing

Hydrogen fuel cells may be developed to supply sufficient power for running vehicles. Hydrogen fuel cells create an electrical charge by separating hydrogen's protons (H^+) from its electrons (e^-). The fuel cell produces only water as its by-product and does not produce greenhouse gas emissions.

speed with which enzyme-catalyzed reactions run, biological fuel cells take time to create enough energy to run a car. So far, biological fuel cells have been employed only for running low-energy devices such as calculators.

Future fuel cell vehicles may circumvent the problem of low power production by incorporating stacked fuel cells. Stacked cells contain many single fuel cells lined up in a series to increase the overall voltage they produce. As with any type of vehicle energy source, fuel cells must supply both power and longevity or consumers may not be enticed to try them.

NUCLEAR FISSION AND FUSION

Of all energy sources, nuclear energy supplies the most power and longevity for a unit of fuel. However, nuclear energy has not been considered for passenger vehicles, mainly because the public would have justifiable concerns over their safety and the *radioactive* wastes they would produce. Naval ships have used nuclear power for a long time and have thus conserved billions of gallons of crude oil.

Nuclear fission is a change in an atom's nucleus in which the nucleus splits apart to form new, lighter nuclei. Each fission reaction releases uncharged atomic particles called neutrons plus energy. As the released neutrons crash into other atoms, more nuclei break apart and release more neutrons and more energy. These multiple self-sustaining fissions are collectively called a nuclear chain reaction. Nuclear chain reactions release enormous amounts of energy that must be controlled with utmost care.

Nuclear fusion is the opposite process to fission. In nuclear fusion, two atoms of an element, such as hydrogen, are forced together at high temperatures until they form a new, heavy nucleus. This step also releases a great amount of energy.

Conventional nuclear reactors that produce energy for communities or for powering the massive ships of the U.S. Navy use nuclear fission reactions. Uranium and plutonium serve as the fuels for these reactors. Although nuclear reactions would never be practical for powering vehicles, hydrogen fuel cells offer a type of nuclear reaction that produces safe energy with a viable future.

The Nuclear Energy Institute in Washington, D.C., has proposed that nuclear power may best serve transportation as a major power source for plug-in electric vehicles. Tomorrow's alternative vehicles and nuclear power might work in complementary fashion toward two ends. First, this system would spare fossil fuels by eliminating the gasoline-fed combustion or diesel engines and, second, the system would bypass the need for coal-fired power plants to make electricity. In short-term planning, however, nuclear power remains most realistic for oceangoing ships.

NATURAL GAS FUELS

Natural gas contains 50 to 90 percent methane (CH_4) by volume. These reserves usually occur above crude oil reserves because the processes that formed oil millions of years ago by the breakdown of organic matter also gave off gas as a by-product. Such natural gas reserves located near crude oil reserves are termed conventional natural gas. Some rarer natural gas reserves have been found with no crude oil nearby. These reserves are called unconventional natural gas.

The largest volume of natural gas exists in the Middle East, followed by Russia and its neighboring countries. The United States owns about 3 percent of the world's natural gas reserves.

Natural gas as an energy source holds many of the same drawbacks as crude oil. That is, natural gas is a nonrenewable resource that will eventually run out at its current rate of consumption. The total known reserves of natural gas should last for another 200 years at the present rate of consumption, but this is only an estimate, and it may be an optimistic estimate. That is because some of the natural gas yet to be tapped consists of unconventional natural gas that is very costly to extract from the Earth. For this reason unconventional natural gas reserves have remained largely untouched.

Like crude oil, natural gas requires an energy-demanding series of steps to get it out of the ground and turn it into usable fuel, shown in the following steps:

1. exploration—searching for reserves by *geological surveys*, mapping them, and setting up drilling plans
2. extraction—building drilling wells and getting the gas out of the ground
3. production—separating various hydrocarbons from the methane that serves as a fuel source
4. transport—delivery of the gas through an extensive network of underground pipelines throughout North America

5. storage—holding in underground tanks until needed
6. distribution—delivering quantities to consumers such as households and businesses

The steps listed here are not the same for every natural gas reserve. For instance, sometimes exploration requires more than one technology to find a natural gas deposit. Exploration teams supplement geological surveys with seismology, which studies the density of the Earth's layers by pulsing energy waves underground. Seismologists create three-dimensional images from their data to picture all underground formations. Other instruments measure subsurface magnetic fields or gravitational fields. Even with the best evidence, companies must sometimes drill exploratory wells to probe for natural gas. In addition to these considerations, drilled wells behave differently depending on how close they lie to an oil reserve.

Once the gas has been drawn aboveground, processors adjust their production method to two components of the gas: the blend of hydrocarbons present and a component called natural gas liquids. Natural gas liquids consist of ethane (two carbons), propane (three carbons), butane (four carbons), and natural gasoline. The natural gas industry sells each of these components as separate products for distinct uses—propane heats homes; handheld lighters use butane.

Despite the difficulties in finding and extracting fossil fuels, the Energy Information Agency (EIA) expects natural gas demand to increase for residential and commercial use and electricity production by 2025. The EIA reports that industry currently runs its operations with the following fuels: petroleum, 45 percent of energy use; natural gas, 37 percent; coal, 9 percent; and renewable sources, 9 percent. Unless new sustainable technology becomes commonplace for industry and the public, the EIA predicts this same usage through 2025.

Natural gas supplies very little energy to transportation, but proponents of natural gas vehicles feel that this energy source can play a role in conserving oil. Natural gas produces energy in combustion similar to gasoline, so technology already exists for using it as a fuel. Early attempts at using natural gas to power vehicles showed that very large fuel tanks would be needed. Scientists therefore worked on new types of natural gas, treated to take up less space. Liquefied natural gas (LNG) has been cooled to a temperature that makes it a liquid that fills less volume than gas.

Compressed natural gas (CNG) forms when natural gas is stored under pressure to reduce its volume.

Currently, CNG cars outnumber LNG cars, but increased sales of any natural gas vehicles have been deterred by the very limited availability of natural gas pumps. Natural gas vehicles also have a shorter range at present than gasoline-powered vehicles.

Biologically formed natural gas, called biogas, consists of methane minus the natural gas liquids. (The term *biological* may be misleading since regular natural gas also formed through biological means over eons.) Biogas arises from the fermentation of organic matter by microbes. Manure, landfills, and wastewater treatment plants all produce biogas that can be collected and used in the same way as natural gas.

Waste Management, Inc., North America's largest waste hauler, has been developing a process for capturing landfill methane and converting it to fuel for its trucks. The company intends to use technology developed in Germany to purify the methane, then chill it to –260°F (–163°C) so that the gas turns to liquid. Company vice president Kent Stoddard said in 2008, "We're creating a valuable resource at our landfills." Landfills and wastewater treatment plants therefore offer an unlikely but possible fuel source for a new generation of vehicles.

NEXT GENERATION HYBRIDS

In the near future, hybrids will assuredly lead the alternative fuel vehicle market. Electric-gasoline hybrids allow car manufacturers to retain the combustion engine with a fairly minor revision by adding a battery power source as in the Prius. Oil companies can furthermore continue to sell gasoline at their current pumps. But change will come to automakers and oil producers nonetheless. Hybrids such as the Prius travel 45 or more miles on a gallon [19 km/l] of gasoline, which conventional cars cannot match. Americans and drivers in other countries have shown an increasing desire to buy personal vehicles that help conserve nonrenewable fuels, especially when gasoline prices rise.

Automotive engineers face a variety of tasks for improving hybrid cars to increase their appeal. New hybrids will likely contain battery-gasoline or fuel cell–gasoline models or the same models with natural gas or biogas substituting for gasoline. All of these models will need readily available fueling stations, although the number of stations could be reduced because they will set better fuel mileage. Batteries must continue to be made lighter and be able to retain a charge much longer than current batteries.

Flex-fuel cars also offer an option in hybrid technology in which the vehicle's engine can run on any gasoline-ethanol mixture from 0 percent ethanol to 85 percent ethanol. The EPA has estimated that more than 6 million flex-fuel cars now travel U.S. roads. The Big Three automakers sell about 40 different models in the United States—Europe and Brazil also sell a large number of models. The main producers of flex-fuel cars other than the Big Three are Citroën, Fiat, Honda, Isuzu, Koenigsegg, Mazda, Mercedes-Benz, Mitsubishi, Nissan, Peugeot, Renault, Saab, Toyota, Volkswagen, and Volvo.

Though some automakers seem to have been slow in making vehicles that do not depend solely on fossil fuel, the industry has been working to make up for lost time. At the 2009 International Auto Show in Detroit, the Michelin company announced the Michelin Challenge Design competition for 2010: "At a time when vehicles are becoming more fuel-efficient and automotive manufacturers are tasked with bringing consumer-friendly vehicles to market that will contribute to environmental sustainability, Michelin appropriately announces its theme for the 2010 vehicle global

Prototype vehicles give the automobile industry and the public an idea of the appearance of new models for the future. This Aptera 2h is a diesel-electric hybrid. Buyers may preorder a 2h, which the manufacturer plans to begin selling in 2010. *(Aptera)*

design competition, Michelin Challenge Design (MCD) as: 'Electrifying! Beautiful, Innovative and Radiant.'" Michelin's tagline perhaps unwittingly shows the obstacles that a new generation of cars face—the 2010 competition emphasized electric power, but the car's appearance seems to be of equal importance. Any new technology, no matter how good it is for the environment, must satisfy the tastes of car buyers.

CONCLUSION

Cars and trucks contribute significantly to the air pollution that causes climate change, so new vehicle and fuel technologies represent two of the most important areas in reducing humanity's ecological footprint. These new technologies must deliver changes rapidly because vehicle sales have been increasing steadily, especially in countries that have both high populations and robust economic growth. The transportation industry must make up for lost time in finding cleaner alternatives to the world's current fleet of vehicles.

The types of alternative fuel vehicles that succeed in the long term will be determined by the new nonfossil fuels that technology develops. Early attempts at ethanol fuel from corn seemed to be a perfect answer for a time until economists and human aid organizations began to see troubling changes in world food supplies. Governments, agriculture policymakers, and free trade markets must find a way to supply new biofuels without interfering with world food production.

The future of alternative fuels will likely be biofuels made from a variety of crops or from biomass, natural gas processed in a way that makes it practical for personal vehicles, and a new generation of fuel cells to replace batteries. Hydrogen fuel cells already show promise and entrepreneurs are also investigating biological fuel cells as an answer to clean, efficient power. All of these plans depend to a great deal on the willingness of automakers to design cars to run on alternative fuels.

Perhaps a new generation of vehicles to replace the century-old dependence on gasoline vehicles requires more than technology alone. Because drivers in industrialized countries have a strong attachment to their vehicles, the automotive industry knows it must satisfy car buyers' tastes at the same time it introduces clean-vehicle technologies. With support from government leaders, economists, and environmental scientists, the transportation industry can launch a new era in road travel.

BIOREFINERIES

Biorefining refers to the production of liquid fuel from plant constituents. Biorefining technology has emerged as a priority in environmental science for a critical reason: humanity's insatiable thirst for crude oil. The world's oil companies produce about 83 million barrels of oil a day, but the amount of oil the world consumes each day totals 1 to 2 million barrels more than current production rates. (A barrel contains 42 gallons [159 l] of oil.)

In 2004, the *National Geographic* writer Tim Appenzeller noted, "Humanity's way of life is on a collision course with geology—with the stark fact that the Earth holds a finite supply of oil." Although the Earth still holds large crude oil reserves under continents and oceans, the oil industry has reached a point in which all the easy-to-reach oil has been drilled. Each new oil extraction has become increasingly difficult and expensive to execute. Oil experts vary in their opinion of how much oil remains to be extracted from the Earth. If crude oil production continues at its current global pace, the British oil expert Colin Campbell, working with the U.S. Geological Survey (USGS), has predicted the peak oil supply will be reached between 2016 and 2040. The Saudi Arabian oil geologist Sadad al Husseini has calculated that the peak may arrive even sooner. The world's peak oil supply is a critical factor for the future because after it has peaked, the increasing supply of cheap oil changes to a declining supply of expensive oil.

The geophysicist M. King Hubbert proposed as early as 1949 the idea of the world oil supply peaking within this generation's lifetime. In an article "Energy from Fossil Fuels," Hubbert described the situation in coal, petroleum, and natural gas reserves that the world faced since

Azerbaijan ranks 22nd in world oil production (1,099,000 barrels per day), with significant drilling in the Caspian Sea. This oil field at Azeri has been depleted. Depleted oil fields can be either plugged and the operations shut down, or they can be modified as natural gas storage sites.

the start of the 20th century when the human population began to grow faster than in any other time in history: ". . . the events which we are witnessing and experiencing, far from 'normal,' are the most abnormal and anomalous in the history of the world. Yet we cannot turn back; neither can we consolidate our gains and remain where we are. In fact, we have no choice but to proceed into a future which we may be assured will differ markedly from anything we have experienced thus far." Many environmentalists have heeded Hubbert's chilling prediction, but humans in general do not take lifesaving steps until they are forced to do so. Someday biorefining will be viewed as being as critical to the world's progress as oil refining is today.

The United States needs new energy technologies for reasons in addition to the inevitable peak in world oil supply. U.S. leaders have worried over the fact that the largest crude oil reserves in the world are in the Middle East, which is a worrisome political region. The United States can go in either of two directions to relieve the coming oil demand crisis: It can tap new domestic oil reserves or it can make significant progress in new technologies for alternative fuels. Biorefining belongs to the second option. Biorefining produces an alternative fuel from either solid biomass or liquid oil or grease wastes. This chapter discusses the status of biorefining in relation to conventional oil refining. It describes today's refining industry and explores the aspects that refineries must consider in order to convert their operations to biomass refining. The chapter also covers topics of the refining industry that have affected how oil refining operates and may influence future biorefining. These topics are pipeline management, the U.S. Department of Energy (DOE), and the economics of oil.

TODAY'S REFINERY INDUSTRY

The present global oil refining industry produces fuel for cars and trucks, buses, aircraft, and ships as well as for non-vehicle uses such as road asphalt components, home heating fuel, lubricants, raw materials for plastics, and petrochemicals. The demand for all oil products has risen with increased global business, which demands more fuel for intercontinental transport, and a growing population.

Increasing demand for oil has helped build the oil industry into the most dominant industry in the world. Because this industry's profits derive from a nonrenewable resource that will become increasingly difficult to extract from the Earth, oil's price per barrel can be expected to steadily increase. If the price of oil exceeds the ability of businesses and the public to buy petroleum products, the oil industry will not be able to sustain its current way of doing business.

In addition to oil reserves that are more difficult to find and extract, oil-refining technology shows signs of difficulty in keeping up with the world's demand for oil. Husseini explained in a 2005 interview with Steve Andrews of the Association for the Study of Peak Oil (ASPO), "Oil capacity today is not production limited but rather processing limited. That is to say, the DOE reports the world's refining capacity has leveled at around 83 MMbd [million barrels of oil per day] for some time and refinery expansions are slow and costly." For this reason Husseini predicted "oil production will level off at around 90–95 MMbd by 2015. A rapid global refinery expansion program that eventually matches an increasing oil demand rate of 1.5 to 2 percent per year cannot be achieved before 2015 at the earliest and is highly improbable in any case." Al Husseini continues to stand by his projections, and many energy experts are listening.

The United States has already entered an oil-deficit situation. The country reached its peak oil production in the 1970s. Since the 1940s when world oil production averaged almost 200,000 barrels a day, the average U.S. daily oil production has declined each decade until it reached 77,000 barrels a day during the 1990s. Even nonscientists should realize that oil production could drop even further during the 2000s.

Because of the U.S. oil production declines, leaders in government have proposed opening up new untapped oil reserves now lying under protected land, such as national parks, wilderness areas, and marine protected areas. In 2008, George W. Bush called the restrictions on offshore oil drilling "outdated and counterproductive." Governor Arnold Schwarzenegger of California, where much of the offshore oil lies, agreed and added, "We are in this situation because of our dependence on traditional petroleum-based oil." But the governor has suggested a different tactic to more drilling

THE MAIN ALTERNATIVE FUELS FOR VEHICLE USE		
FUEL	**SOURCE**	**DESCRIPTION**
biofuels	vegetable oils and animal fats from the food industry	replaces either gasoline or diesel and less polluting than standard fuels
ethanol	corn and other crops	plentiful and produces few greenhouse gases
hydrogen	coal, nuclear power, renewable sources such as hydropower, or fuel cells	emits only water and nonhazardous gases
natural gas	fossil fuel often found near oil reserves	less air pollution and greenhouse gases than gasoline or diesel
propane (liquefied petroleum gas)	crude oil refining	less air pollution and greenhouse gases than gasoline or diesel

Coal accounts for 27 percent of world energy consumption, and the U.S. Energy Information Administration (EIA) expects that share to grow to 29 percent by 2030. Asia consumes more coal than any other world region, which explains in part the pollution problems that occur in many industrial Asian cities. Although coal mining and burning creates pollution, countries have depended on them because coal is plentiful and inexpensive. The future of coal energy depends on new clean-burning technologies and coal-to-fuel conversion. *(Tom Weiland)*

by advocating "new technologies and new fuel choices for consumers." The responsibility falls squarely onto the shoulders of the refining industry.

The refining industry's future is not coming to an end. The world's oil reserves have not yet run dry, but nonetheless they are gradually disappearing. In light of the predictions for future oil production, refineries must begin now to plan and retool for alternative fuels. The refining industry has the opportunity to attack the fuel problem from two directions: biorefining and innovations in traditional oil refining. Both of these approaches will seek the goal of producing fuels no longer based on crude oil. The alternative fuels described in the table on page 78 have the potential to supersede crude oil in this century.

As the world approaches its peak oil production, oil industry researchers have begun investigating options in addition to biofuels and the alternative energy sources listed in the table. Two long-range approaches consist of coal-to-oil processing and tar sand processing. In the coal-to-oil method, coal—still an abundant resource—is turned into a liquid fuel under high temperature and pressure. But some coal-to-oil drawbacks include high costs and large amounts of exhaust emissions. Large quantities of sand in western Canada hold tar, which is a substance from crude oil that has migrated toward the Earth's surface. Tar extracted from Canada's sand reserves can be turned into crude oil, but this process demands large amounts of water and energy. Both coal-to-oil and tar extraction will demand long-term research. They will likely not solve the oil problem, but may someday serve as a supplement to dwindling oil reserves. The biorefining industry hopes to develop new alternative fuels before drivers will ever need to depend on coal or tar.

Disasters such as 2010's Gulf of Mexico oil drilling platform explosion and spill have put tremendous pressure on oil companines. The industry faces a new reality for meeting global fuel demands in an environmentally safe manner. The sidebar on page 81, "The U.S. Department of Energy" discusses how this government agency makes decisions on the energy future of the United States.

PIPELINES

Pipelines in the United States and almost every other part of the world carry crude oil from drilling sites to tankers or straight to oil refineries or carry natural gas to gas refineries. Several thousand oil tankers transport billions of gallons of oil across the oceans each day to carry oil where pipelines do not exist. Ships and pipelines receive unwanted publicity whenever an accident or leak occurs, but in light of the massive amount of oil they transport through rough seas and over remote terrain, oil transport has been a generally safe activity.

In the United States, the Trans-Alaska Pipeline System (TAPS) stretches for 808 miles (1,300 km) from Prudhoe Bay on Alaska's North Slope near the Arctic Circle to the Port of Valdez in the south. Six different pipeline companies run the line known by Alaskans as the Alyeska pipeline, and along its route the pipeline bisects the largest expanse of unspoiled wilderness in the United States.

Because of the Alyeska pipeline's remote path through pristine wilderness, environmentalists voiced concern during the pipeline's planning

THE U.S. DEPARTMENT OF ENERGY

The DOE began operations in 1977, five years after Jimmy Carter proposed that a single department within the president's cabinet should administer the country's energy policy. The department now numbers about 14,000 employees with its main office based in Maryland near Washington, D.C. Under the leadership of the secretary of energy, the DOE has three main responsibilities: (1) coordinate the nation's energy supply and use; (2) lead research and development in energy conservation and alternative energy sources; and (3) supervise the nation's production and disposal of nuclear weapons.

The DOE's science and technology program funds research in a variety of areas, many of which focus on climate change and issues related to global warming, such as greenhouse gases. The department is also occupied with solving the problem of U.S. oil production, that is, that production is decreasing but demand is increasing.

The DOE investigates biological and other renewable energy sources for the purpose of helping to ease the discrepancy between oil use and production. Part of the research on biomass fuels concentrates on achieving the most efficient ways to convert biomass to liquid fuels. The DOE has suggested that testing should be done not only on biomass to make hydrocarbon fuels to mimic petroleum, but also microbial sources, thermochemical reactions that combine chemical conversions under high heat—coal-to-oil offers an example—or advanced chemical methods using catalysts. As part of the research it supports, the DOE has emphasized that current refining technology must expand into these new areas.

In 2009, President Barack Obama named the Stanford University physicist Steven Chu as the new secretary of energy. As the DOE has done under other administrations, it will be expected to implement the president's national energy policy. New national energy policies will emphasize renewable energy sources, and the DOE will be expected to take the lead in alternative and renewable energy technologies. Chu told the *Washington Post* shortly after his appointment, "I was following [climate change] just as a citizen and getting increasingly alarmed. Many of our best scientists now realize that this is getting down to a crisis situation." The DOE will soon put renewed efforts into solving climate change with an urgency never before seen in previous administrations.

The Trans-Alaska Pipeline at Kuparuk on the Alaskan Arctic plain. In 2006, the pipeline ruptured and spilled more than 270,000 gallons (1 million l) of oil into Prudhoe Bay. Cleanup teams struggled in temperatures of −63°F (−52.8°C) to dislodge oil frozen to ice. The spill rekindled debate over proposed drilling in Alaska's Arctic National Wildlife Refuge (ANWR). *(William Breck Bowden)*

and building and the controversy over the pipeline continues four decades later. In 2009, an *Audubon* magazine article stated, "Since the discovery of oil in Prudhoe Bay in 1968, development has spread east, west, and offshore, sending several billion barrels of oil south through the Trans-Alaska Pipeline. The 19 producing oil fields on the North Slope are spread over 1,000 square miles [2,590 km²] of *tundra* and wetland. Roads, pipelines, drilling pads, airstrips and other infrastructure in the central Arctic oilfields have covered more than 9,000 acres [2,590 km²] of tundra." The Alyeska's proponents have, of course, argued the benefits of the project, discussed in the sidebar "Case Study: Alaska's Oil Economy" on page 84.

The Trans-Alaska Pipeline is far from the longest—or even the most controversial—of the world's oil pipelines. The Druzhba pipeline, the world's longest, covers 2,500 miles (4,000 km) from central Russia to eastern Europe and Germany. The second longest line, the Baku-Tbilisi-Ceyhan pipeline, stretches less than half the distance (1,099 miles; [1,768 km]) from the Caspian Sea to the Mediterranean Sea. The capacity of the world's largest pipelines may be difficult to determine because countries and oil companies often keep the details of their pipelines from the public. They hold information on new construction and changing capacity of pipelines in check for the following reasons: (1) security against terrorist attack; (2) as a business decision in response to fluctuating world oil supplies and prices; and (3) as a protection against environmentalists who may oppose the building of new pipelines.

The natural gas pipeline network in the United States also represents one of engineering's most impressive accomplishments. Natural gas from the main reserves in the Texas-Louisiana Gulf of Mexico region, Oklahoma, western Texas, and West Virginia-Ohio-western Pennsylvania travels in underground pipelines to each of the 48 contiguous states. These pipelines lead to large underground storage facilities located in 28 states. The United States uses three main storage methods for natural gas: (1) depleted natural gas or oil fields; (2) salt caverns; or (3) natural underground water reserve sites called aquifers. In a few rare cases, abandoned coal mines have also been converted into gas storage facilities.

Each type of natural gas storage presents advantages and disadvantages. Salt caverns have been used increasingly since the 1980s because of their stability and ease of injecting and removing the gas. Aquifers, however, require extra preparation to assure that deep rock formations hold the gas without contaminating nearby underground water sources. Each of the current methods for natural gas storage requires expenses that become part of the overall price that consumers pay for natural gas.

Pipelines crossing international borders and covering very long distances present increased chances for accidents and spills or possible disruption due to political actions. National energy commissions therefore monitor pipelines for the following events to assure pipeline safety:

- breaks and leaks due to aging
- damage due to earthquakes, floods, freezing, or storms
- war or conflicts near pipelines
- interference at border crossings between neighboring countries

Environmental scientists have shown another very serious concern regarding pipelines—their interference with wildlife migrations. Alaska's caribou carry out one of the world's largest wildlife migrations, and, early in the Trans-Alaska Pipeline planning, environmentalists feared that

CASE STUDY: ALASKA'S OIL ECONOMY

Alaska's economy depends in part on the robustness of the world oil market. Alaska produces about 10 percent of the oil used in the United States and also exports oil to countries in Asia; the state produces more than 20 million barrels of oil per month. For this reason, Alaska balances two often-competing interests: Alaska holds immense oil reserves needed by the United States, but it also contains the nation's largest swath of undisturbed wilderness located on or near additional oil reserves. Can Alaska reconcile oil production with environmental protection?

Alaska receives most of its income from the following industries: oil and natural gas, timber, fishing, mining, and tourism. According to the state chamber of commerce, the oil and gas industry generates almost 34,000 jobs, and taxes on oil companies' revenues brought the state more than $10 billion in 2008, double the amount of the year before. Each Alaskan receives a yearly check of between $1,000 and $2,000 from the state's savings account from money earned by investing oil tax revenues.

Because of the wealth that Alaska receives from oil, residents and government leaders may be tempted to raise the taxes it levies on oil companies drilling in the state. While the tax income benefits Alaska and its citizens, Alaska must also be careful not to tax companies to a level that makes further exploration and drilling too expensive. Marilyn Crockett of the Alaska Oil and Gas Association, which represents oil and gas companies, said in the *Seattle Times* in 2008, "Clearly, from the investor standpoint, Alaska has become a less attractive place to invest exploration and production dollars." In short, Alaska wants and possibly needs the oil industry and must make decisions to support oil drilling.

Some Alaskans have become such strong proponents of fossil fuels that they question whether global warming truly exists. The columnist Dan Fagan of the *Alaska Standard* has used such terminology as the "global warming cult" and refers to questions about climate change as "hysteria." Nevertheless, many Alaskans worry about activities that encroach onto unspoiled lands. The Alaska Department of Commerce cites as its mission the need to maintain existing energy programs but also to pursue new energies that are "sustainable and environmentally sound." This department now supports programs in fuel efficiency, biomass energy, geothermal energy, hydroelectric energy, ocean and river energy, solar energy, and wind energy.

construction and the pipeline itself would alter the natural migration route and permanently damage the tundra ecosystem. To reduce the potential harm to wildlife, the pipeline contains 23 buried sections and more than 500 elevated (10 feet; [3.3 m]) sections to allow caribou and other wildlife

The Trans-Alaska Pipeline crosses the Brooks, Alaska, and Chugach Mountain Ranges on its route between Prudhoe Bay's oil fields and the tankers waiting at Port Valdez. The pipeline symbolizes the strong relationship between Alaska's residents and oil. The Alaskan economy depends to a large extent on its natural resources: fishing, timber, wildlife, crude oil, and minerals. *(Southwest Research Institute)*

Alaska's alternative energy programs are in an early stage. Former Alaska senator Ted Stevens cautioned in 2008, "Alternative energy opportunities in Alaska are enormous, but it takes a large investment to get them started." Meanwhile small local groups have discussed plans for investigating alternative energy, and some entrepreneurs have opened stores that sell devices for maximizing fuel efficiency and converting to greener methods. For the present, however, many Alaskans see a more lucrative future in oil than in renewable energy.

to pass, but environmentalists clearly wish the pipeline had never been built. *Audubon* noted in 2009, "In the early 1990s the Alaska Department of Fish and Game found that caribou inhabiting the oil field had lower calf productivity than animals from the same herd that calved farther away from oilfield-related facilities." Other mishaps have confirmed environmentalists' worries. Leaks and spills and electrical and mechanical breakdowns at pump stations have occurred on occasion.

Monitors of the Alyeska pipeline try to assure that any spills or leaks are contained quickly. Meanwhile, environmental scientists continue to study the effects the pipeline has had on wildlife. Unfortunately, less may be known about the environmental effects of pipelines that cross Russia, Siberia, and eastern Europe's forests.

Pipelines might be necessary for any type of fuel in the future, including nonfossil fuels. For this reason, pipeline planners and engineers must work with ecologists to develop low-impact structures that serve people and protect nature. Just as important, safeguards for the environment must be in place throughout the world's pipelines.

BIOREFINING STEPS

Biorefining refers to the production of fuels using biological materials with the intention of replacing gasoline or diesel. The biological materials are made up of solid biomass or vegetable-based oils. The biorefining industry has also investigated technologies that start with fossil fuels, but in a more efficient and nonpolluting process than conventional refining.

Biomass made of composted organic material or animal wastes can be turned to a thick, dark fluid that the biorefining industry calls bio-oil. Biorefiners produce bio-oil by heating biomass until it decomposes into constituent hydrocarbons. This heating process called pyrolysis can also be used for making natural gas. Biorefiners usually employ flash pyrolysis to make fuel, so-called because the process heats biomass to a very high temperature in a short period of time. The heating step emits vapors that the biorefiner condenses to an oily collection of hydrocarbons similar to those in crude oil. Subsequent refining steps then mirror conventional oil refining. To date, bio-oil works best as heating fuel and as a source of industrial petrochemicals. Biorefiners also capture the gas given off in the production of bio-oil and further refine it to make an additional energy source.

Vegetable oils from soybeans, cottonseeds, palm, sunflower, jatropha, and frying oils or grease wastes from restaurants have been experimented with in modern cars and buses since the 1980s. Oil-processing chambers called bioreactors make about 50,000 gallons [189,270 l] of oil that contains the same mixture of hydrocarbons as required by the government in order to be regulated as a vehicle fuel. Heat and the feedstock oil go into the process, and then propane, a small amount of water, and the liquid fuel exit. In order to get the most out of this process, biorefiners make the thick vegetable oils or greases easier to handle and to turn into fuel. They do this either by diluting the oils or grease with a thinner oil, or they treat the thick substance by a chemical process that changes the hydrocarbon bonds and turns the material into a more liquid consistency. Following these steps, the new fuel has been modified into a form suitable for use in combustion engines.

Biodiesel fuel differs from other biofuels by its hydrocarbon composition. Biodiesel belongs to two different groups based on source and process method. First, pure biodiesel, called B100, comes from biomass and works only in diesel engines. Second, biodiesel blends contain some petroleum mixed in with the biodiesel so that the fuel can be used in a variety of engines. This second type of fuel comes in a variety of biodiesel-petroleum blends; B5 is 5 percent biodiesel and 95 percent petroleum, B20

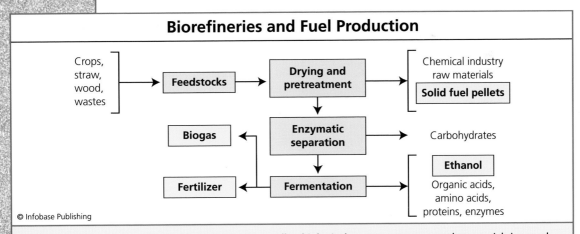

Biorefineries and Fuel Production

© Infobase Publishing

Biorefineries rely on raw materials made by nature as well as biological processes to convert the materials into end products. The biorefining industry has the capability to produce three different types of renewable fuels, highlighted in the diagram: solid fuel pellets for use in biomass energy production; ethanol biofuel; and biogas, a heating fuel. Biorefining also produces chemical-free fertilizers for agriculture and other organic compounds.

is 20 percent biodiesel, and so on. The EPA permits all biodiesel blends to be used in conventional diesel engines.

The main step in biodiesel production comprises the *transesterification* reaction in which a catalyst helps change the structure of the fats in the oil. Biodiesel refineries usually rely on the following feedstock oils: soybean, sunflower, canola, or used fryer oil from restaurants, or restaurant grease or tallow. Transesterification produces biodiesel plus glycerin as a by-product. Glycerin can be refined to make methanol, an alcohol that biorefineries then use as an energy source in their own operations. After removing the glycerin, the biorefiner purifies the biodiesel by removing contaminants such as water, unreacted fats, and small amounts of excess glycerin.

Pure biodiesel produces two-thirds less unburned hydrocarbons from engines and almost 50 percent less carbon monoxide and particles. Pure biodiesel and biodiesel blends produce more nitrogen oxides, however, than regular fuels. Biodiesel's other drawbacks include limited availability and the large area of land needed to support biodiesel manufacture.

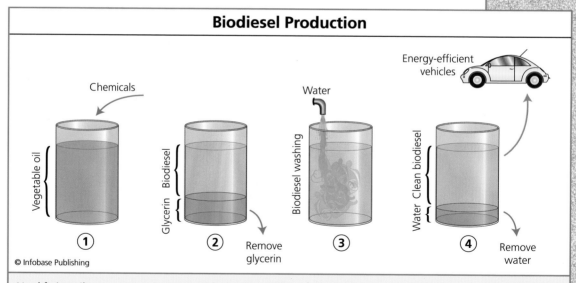

Biodiesel Production

© Infobase Publishing

Used frying oils serve as starting material for biodiesel. The chemicals lye and methanol added to vegetable oil break the fats into biodiesel precursor and glycerin. By removing the glycerin and water-washing impurities from the remaining liquid, a biorefinery produces clean biodiesel. Biodiesel usually requires a final step, called drying, in which the fuel is stored for a period to let the last remaining impurities settle out by gravity. Then it is ready for use in cars and other vehicles. (*Source: John Blanchard*, San Francisco Chronicle)

Drivers of biodiesel vehicles have been frustrated by the lack of fueling stations that cater to them even in environmentally conscious areas such as California. An Oakland, California, resident Jonathan Austin told the *San Francisco Chronicle* in 2007, "You can't just run down to the gas station. You've got to plan ahead." The number of biodiesel pumps has increased only slightly since Austin voiced his complaint. Meanwhile, biorefineries have pursued feedstocks that produce the most fuel from a given area of land. Canola oil can produce up to 150 gallons [568 l] of biodiesel per acre [0.004 km²] of land planted with rapeseed; sunflowers produce about 100 gallons [378 l] per acre; and soybeans make 50 gallons [189 l] of fuel per acre.

Biorefineries have not reached the number and scale of today's massive oil refineries. Biorefining has largely confined itself to ethanol fuels for the past several years, but disadvantages in large-scale ethanol production have prompted the biorefining industry to investigate new avenues in fuel-making. The DOE and government leaders will be called upon very soon to create a clear plan for alternative fuels that benefit the environment, the economy, and a new fleet of vehicles.

DEVELOPING THE BIOREFINING INDUSTRY

The biorefining industry remains young and small, as noted by the *San Francisco Chronicle* reporter Michael Cabanatuan in 2007: "Biodiesel has been popular for years among farmers in the Midwest and in the South, where virgin soybean oil typically is used to produce the fuel. Yet its use in the West, until recently, was largely limited to hobbyists who brewed the fuel at home and people who prided themselves on not using oil." Home production will not be the route to alternative fuels; the EPA does not approve home formulas for use on public roads for two main reasons. First, home-prepared fuels would likely not conform to standard formulas so that the EPA could not estimate their effects on air pollution. Second, home manufacture of fuel is extremely dangerous due to the risk of explosion.

Millions of dollars from investors and the DOE have fed projects focused on the expansion of biodiesel refining capabilities. Larry Russo of the DOE said, "We need to do the research of course, but then we need to do the pilot testing with our partners, and then scale these things up to get

The ethanol biofuel industry in the United States grew primarily due to large harvests of corn, the feedstock for ethanol production. The cornstalks can be used as feedstock for biomass energy production. Concerns over the effects of massive corn harvests dedicated to ethanol production have prompted some farmers and economists to suggest different sources for ethanol, such as switchgrass, which requires less energy input to grow than corn.

to the point where it can attract financing on its own." Pilot plants carry out all the steps of a full-scale manufacturing plant, but the pilot plant is a much smaller facility designed to debug the system and look for ways to improve efficiency.

In May 2008, Congress passed the Food, Conservation, and Energy Act, known familiarly, as the 2008 Farm Bill. This act stipulated the availability of more than $1 billion for diverse programs, including biofuel, biomass, and technologies to expand biorefining. Large oil companies have also invested in biorefining technology centers to speed development. Despite the increased activity in biofuel research, biorefining's future remains largely unknown. Even the most knowledgeable experts in alternative fuels have yet to predict which alternative fuel will become the viable replacement for gasoline and diesel.

One of the problems biorefining must solve comes from the large amounts of glycerin by-product that accumulates in the biofuel-making process. Glycerin poses no hazards, but it has to go somewhere. Some of the excess glycerin (also called glycerol) from biorefining goes to other industries for making soaps, moisturizers, antimicrobial formulas in veterinary medicine, and as a carrier for some human drugs. A chemical engineer Kenneth F. Reardon of Colorado State University told the *New York Times* in 2007, "Just like petroleum refineries make more than one product that are the feedstock for other industries, the same will have to be true for biofuels. Biorefining is what the vision has to look like in the end." Large biorefineries take a long time to perfect and build, and part of their planning must include a new use for glycerin. These represent big challenges for the biorefining industry. No one is sure for the moment whether biorefining can overcome its obstacles in a timely fashion and make an important contribution to the world's alternative fuel supply.

CONCLUSION

Biorefineries play a central role in assuring the success of new alternative fuel vehicles. Unless alternative fuels become as available as gasoline is today, their future is questionable. A large responsibility rests on engineers to design biorefineries that can meet the fuel needs of drivers. This is a daunting challenge. Today's oil companies serve as the largest business enterprises in the world, yet even oil production may not be keeping up with the growing demand for fossil fuels. Biorefining therefore faces dual challenges in finding the best technologies for converting crops to fuel and for producing biofuel on a scale that satisfies world fuel consumption.

For the present, biorefining must identify the best feedstocks for making biofuel in a fast and inexpensive manner. Biorefining produces less air pollution than conventional oil refining, but biorefineries face problems of disposing of large amounts by-products made in the biorefining process. Biorefining must additionally work with governments to create a plan for making the new biofuels available and priced for the driving public.

Biorefining's future shares a strong connection with automakers' current attention to new alternative fuel vehicles. No one yet knows whether biofuels will replace fossil fuels or if biofuels will merely serve as a bridge in technologies between crude oil and fuel cells or all-electric vehicles of the future. Because of the complexity of solving the crisis of the world's growing ecological footprint, biorefining will probably play a role in conjunction with many other technologies. It will be up to biorefiners to determine whether the role of biofuels will be major or minor, long term or short lived, in building a sustainable society.

INNOVATIONS IN CLEAN ENERGY

Clean energy refers to forms of energy that do not produce hazardous emissions, do not harm human and ecosystem health, and do not destroy the environment during extraction from the Earth. Coal-burning plants and gasoline-burning vehicles do not meet this definition. Clean energy has become synonymous with renewable energy, but clean energy also might be from fossil fuels if the fuel can be extracted, processed, and burned in ways that do not harm the environment.

The U.S. Environmental Protection Agency (EPA) provides an online calculator to help people understand the cleanness of their energy sources. Such a calculation serves as a good starting point for learning about the impact everyone has on energy consumption and emissions. The calculator factors in the location where a person lives because some parts of the country rely almost entirely on coal-fired power plants for energy, while other regions use mainly hydroelectric power. These two energy sources have different values when calculating clean energy. The calculator estimates the amounts (in pounds) of carbon dioxide (CO_2), sulfur dioxides, and nitrogen oxides produced by individuals according to their zip code and characteristics of their lifestyle. Users of the calculator additionally learn whether their emissions occur at rates higher or lower than the national average. For example, a U.S. resident might produce the following typical annual amounts of greenhouse gases from energy usage:

- CO_2: 1,400 pounds (635 kg) per *megawatt-hour* (MWh)
- sulfur dioxides: 6 pounds (2.7 kg) per MWh
- nitrogen oxides: 3 pounds (1.4 kg) per MWh

Clean energy plays a part in protecting the environment in addition to reducing emissions that cause global warming. Clean energy conserves nonrenewable forms of energy, reduces environmental damage caused by exploration and extraction of fossil fuels, and minimizes exposure of people and wildlife to large energy production plants.

This chapter describes the important clean energies that are becoming more vital every day in efforts to halt global warming. The chapter also describes innovative technologies in clean energy. It opens with a review of how the alternative energy movement became established. The chapter then discusses wind, water, solar, geothermal, nuclear, and fuel cell power. Each discussion includes the advantages and disadvantages of these technologies. The chapter stresses the process of managing carbon, explaining how people's actions inevitably relate in some way to the Earth's *carbon cycle*.

ALTERNATIVE ENERGY EMERGING

Before 1900, horse-drawn vehicles traveled the roads, only forced to step out of the way for a steam-powered contraption on rare occasions. At the beginning of the 20th century, the German engineer Carl Benz began manufacturing vehicles powered completely by gasoline-fed internal combustion engines. The invention produced more power than steam engines and was also easier to navigate. In 1867, the German engineer Nikolaus August Otto developed a four-stroke internal combustion engine for use in manufacturing equipment rather than road vehicles. Benz, Gottlieb Daimler, and Ferdinand Porsche all saw potential in Otto's design, and they worked independently as well as cooperatively to develop the first practical use of gasoline power for vehicles.

Massive reserves of oil that were discovered under continents or the seas led the next two generations of drivers to believe the tap would never run dry for their vehicles or for heat for their homes. But in 1973, the Organization of the Petroleum Exporting Countries (OPEC), an oil-producing cartel (see the following table), raised the price of oil four times. Suddenly, oil-importing countries such as the United States began to think about alternative energy sources on a large scale and for the long term.

Though inventors had already designed vehicles that ran on electricity, batteries, or combinations of gasoline and non-gasoline fuels,

OPEC COUNTRIES		
COUNTRY	**LOCATION**	**DATE JOINED OPEC**
Algeria	Africa	1969
Angola	Africa	2007
Ecuador	South America	1973–1992, 2007
Iran	Middle East	1960
Iraq	Middle East	1960
Kuwait	Middle East	1960
Libya	Africa	1962
Nigeria	Africa	1971
Qatar	Middle East	1961
Saudi Arabia	Middle East	1960
United Arab Emirates	Middle East	1967
Venezuela	South America	1960

Note: OPEC founding members are Iran, Iraq, Kuwait, Saudi Arabia, and Venezuela

oil-importing countries began to see alternative energy as an increasingly critical new industry. In the 1970s, scientists discovered another disquieting fact: Many oil experts estimated that the United States had reached its peak oil production and would thereafter depend on foreign sources to make up the difference to meet its needs. Many of the same scientists have pointed out that the rest of the world is also close to reaching this peak on the *Hubbert Curve,* which is a graphical depiction of oil supply and consumption. The most important feature of the Hubbert Curve is its ability to estimate a point in time in which fossil fuel demand exceeds supply.

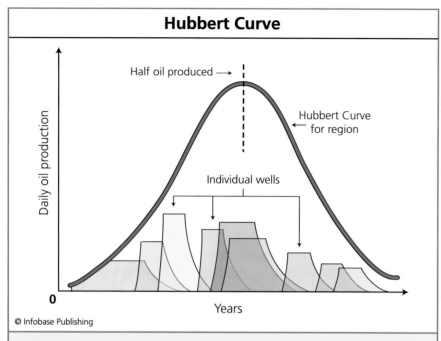

Hubbert Curve

Half oil produced →

Hubbert Curve for region ←

Individual wells

Daily oil production

0

Years

© Infobase Publishing

In 1956, the geologist M. King Hubbert used this graph—known as the Hubbert Curve—to predict when an oil-producing region would reach its peak production. The oil industry dismissed Hubbert's prediction of an end to oil, but in 1970–71 geologists and oil experts calculated that the United States had reached its oil-producing peak. It has used up more than half of its oil production capability. The remainder will be increasingly difficult to find, drill, extract, and produce in a cost-effective manner.

The world's need for energy sources to replace fossil fuels therefore originates from a combination of commerce, world politics, and the Earth's composition; the Earth will not make more fossil fuels in time to meet the human population's increasing consumption rate.

Inventions that were once the domain of amateur scientists have now entered mainstream planning for alternative energies: water, wind, sunlight, geothermal emissions, biological reactions, and hydrogen. In 2007, the *New York Times* reporter Matt Richtel wrote of the new energy platform, "Silicon Valley's dot-com era may be giving way to the watt-com era." Just as computers and the Internet (the dot-com companies) grew from amateur experiments in California's Silicon Valley, the same ingenuity may take the lead in alternative energies.

WIND, WAVE, AND TIDAL POWER

Energy generated from the power of wind, waves, or water uses a passive form of energy collection; the system works by the force of wind or water and does not need any added energy. Wind power has been a fast-growing segment of the alternative energy industry since the 1990s; wind power has grown more than 30 percent in each year of 2006, 2007, and 2008. Wave and tidal power have made smaller contributions, but interest in these modes of producing energy remains steady.

Wind generates power when gusts contact a windmill called a wind turbine. The turbine's blades rotate, which turns gears in a generator behind the blades and inside the turbine. The generator converts the kinetic energy of the rotating blades into electrical energy, which cables then carry to an electrical transfer building to distribute to customers.

Commercial *wind farms* possess hundreds of turbines and usually work best in capturing the wind's energy along coasts or on plains, which receive steady winds in all seasons. Despite this simple solution to producing energy, wind farms have caused some worry in the public's mind. Ernie Corrigan, spokesperson for the Alliance to Protect Nantucket Sound, said in 2006 about a proposed 130-turbine wind farm on Massachusetts's Cape Cod, "Cape Cod has done a very good job of selling this thing as almost bucolic, and it is anything but bucolic. It is an industrial project that would assemble the largest concentration of offshore wind turbines in the world. At night, it would literally transform what is now a crystal-clear skyline into something more like an urban skyline, with tall towers and blinking lights." After more than eight years of lawsuits and government reviews, in 2010 the Barack Obama administration approved the Cape Wind Project, consisting of 130 windmills 440 feet tall (134 m) five miles off the cape's coast.

Does wind power have a future? A 2009 article in *E/The Environmental Magazine* stated, "In 2007, 35 percent of all new electricity generation installed in the U.S.—over 5,200 megawatts—was wind." The article went on to state, "In 2008, wind displaced about 34 million tons [31 million metric tons] of carbon dioxide, equivalent to taking 5.8 million vehicles off the road." Although wind power was expected to produce 30,000 megawatts (MW) of electricity in 2009 and 2010, the U.S. stock market currently contains few publicly traded wind energy companies.

In 2009, General Electric (GE) ventured into wind energy by partnering with China's A-Power Energy Generation Systems to manufacture wind turbine gearboxes. Perhaps GE recognized a growing wind power industry outside the United States. By 2020, wind energy is expected to produce 10 percent of China's energy and take pressure off the enormous amount of coal burned in China. Wind power produces as much as 20 percent of Spain's energy and other countries are approaching that level. The U.S. Department of Energy (DOE) has proposed a similar goal of 20 percent of all energy generated in the United States to come from wind by the year 2030. Wind power has proponents, but as with almost any form of energy it brings disadvantages along with the advantages, shown in the following table.

Renewable wind energy has shown promise as an inexpensive means of producing large amounts of electricity. Wind farms similar to this farm in Indiana take up large land areas along coasts and on ridge tops. Hawks and eagles hunt in those places, and thousands of raptors and migrating birds have been killed by flying into rotating turbines. People living near wind farms have also objected to noise. These drawbacks may be fixable with new technology. *(Indiana Office of Energy and Defense Development)*

WIND, WAVE, AND TIDAL POWER	
ADVANTAGES	DISADVANTAGES
wind	
• efficient converting of wind to electrical energy • moderate to low startup cost • wind is free • no pollution • easy construction • land below wind turbines can be used for other activities	• little power output in low winds • extensive land needed for wind farms • view of wind turbines • injures and kills migratory birds and predatory birds • noise pollution
waves and tides	
• steady source of power • efficient conversion of kinetic energy to electrical power • no pollution • ocean power is free	• useful only on coasts or rivers • unproven technology • expensive construction

Environmentalists have been torn between the benefits of renewable energy from wind and the harm wind turbines cause to migrating flocks of birds and predatory birds such as hawks that hunt across open fields. Some hawks and falcons built nests on early turbine structures, which only increased the chances of adult and fledgling deaths. The wind power industry has tried to mitigate this hazard by developing slower turning turbines and new tubular designs that offer fewer nesting places. The latest turbines to enter the energy market run at about 12 revolutions per minute, which allows birds to see the blades and avoid them.

Wave and tidal power remain unproven technologies for generating useful amounts of energy. The Federal Energy Regulatory Commission (FERC) classifies both of these energies as hydropower (water-generated energy) that makes use of *hydrokinetic energy,* which is energy contained in the movement of water. Several small companies have now

Tidal Energy

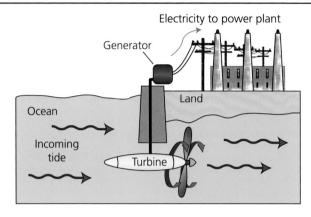

Electricity to power plant

Generator

Land

Ocean

Incoming tide

Turbine

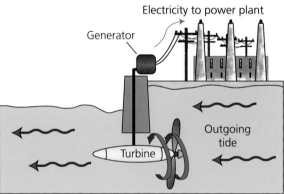

Electricity to power plant

Generator

Outgoing tide

Turbine

© Infobase Publishing

Tidal energy uses the motion of the incoming and outgoing tides to turn turbines and generate electricity. Underwater turbines will be more expensive to install than wind farm turbines, and the turbine blades might injure aquatic species. If those problems can be corrected, tidal energy and related wave energy offer a long-term renewable form of energy.

carved out a business niche by inventing ways to capture wave and tide energy.

Wave power comes from the up-and-down motion of waves about a mile offshore. Various types of floating devices can capture the kinetic energy from wave movements and convert each movement into electricity using a generator. Three main devices for capturing wave energy are the following: (1) pelamis, (2) power buoy, or (3) limpet. A *pelamis* is a large floating chain made of alternating buoys, which float the chain, and power modules equipped with a generator. As the chain undulates over waves, the buoys bump back and forth into the modules, providing enough movement to turn the generators and produce electricity for cables to take to land. A *power buoy* is a single device that bobs on the waves similar to a regular harbor buoy. The constant up-and-down movement runs the generator, and a cable takes the electricity. A *limpet* is an open cylinder on a shore into which waves pound. Each incoming wave rolls into a tube and pushes water against a turbine, which then runs a generator.

Tidal power comes from ocean movements also, but the natural sway of the tides provides the hydrokinetic energy. Underwater turbines rotate with incoming tides and in the opposite direction with outgoing tides. Tidal energy represents a steady and inexpensive energy source, but the costs of building the underwater system can be high. Like wind turbines, undersea turbines may cause harm to marine mammals and fish from the rotating blades. Wave energy and tidal energy have not yet contributed a meaningful amount of energy to power grids, but some proponents anticipate a bright future. "Wind and solar are very diffuse sources—you have to cover a lot of area to collect energy," said Roger Bedard of the Electric Power Research Institute in Palo Alto, California. "Waves carry a lot of energy in a small space. Smaller machines cost less than bigger machines." This type of efficiency will be at a premium as open land fills with new communities each year.

SOLAR POWER

Solar energy produced in the form of light and heat from the Sun help produce heat and electricity in an increasing number of homes, schools, businesses and may some day be a power source for vehicles. Solar energy can be collected by large utility companies that turn it into electricity for their customers, or single buildings can be equipped with a solar thermal system to turn heat into electricity.

Conversion of solar energy to electrical energy depends on a device called the *photovoltaic cell,* also called a solar cell. Photovoltaic cells work by capturing the energy in the Sun's radiation, called photons; the photons then dislodge electrons from a material inside the cell and the flow of electrons produce an electric current. Semiconductor materials such as

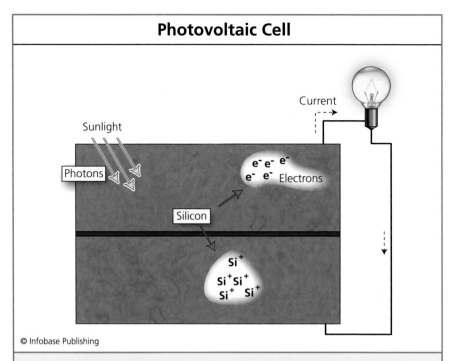

Photovoltaic Cell

Sunlight

Current

Photons

e⁻ e⁻ e⁻
e⁻ e⁻ Electrons

Silicon

Si⁺
Si⁺Si⁺
Si⁺ Si⁺

© Infobase Publishing

A photovoltaic cell used in capturing solar energy receives photons (the Sun's rays), which silicon absorbs. This action releases an electron from a silicon atom each time a photon strikes. Oppositely charged poles on either side of the cell induce the electrons to form a current.

silicon act as the best substance for this conversion of photon energy to electric current.

Solar energy has created the greatest level of interest of all alternative energies for serving homes and other buildings. Solar energy represents a fast-growing industry; most U.S. homeowners live in a place where they can make an appointment and have a sheet of photovoltaic cells, called *solar panels,* installed on their roof in a day. Worldwide, solar installations have increased as much as 62 percent (as MW produced) between 2006 and 2007, yet solar energy accounts for less than 0.05 percent of global energy demand.

The solar energy business appears to hold the most promise of all alternative energies, yet it must overcome its unique hurdles to truly compete with oil (37 percent of total energy use), coal (25 percent), and natural gas (23 percent). Like wind and ocean energy, solar energy collection requires methods of storing the energy until it is needed. Electricity can be difficult to store; batteries store small amounts but cannot yet store the large amounts needed by electric utility companies. Companies such as Ausra in California have explored the idea of storing solar energy as heat, called solar thermal, rather than electricity because this method is less costly than electrical storage. Ausra's vice president John S. O'Donnell explained to the *New York Times* in 2008 that a $5 coffee thermos and a $150 computer battery store about the same amount of energy. "That's why solar thermal is going to be the dominant form," he said. The following table summarizes the current advantages and disadvantages of solar energy.

U.S. leaders and foreign governments have encouraged residences and businesses to increase use of solar energy. Individual towns and entire states have developed solar energy goals. In the United States, some power utilities allow owners of solar homes to receive payment for the energy they do not use. The unused amount then becomes available for others to draw from the community's power grid. Even in towns where installation would be expensive and require decades to make up the cost in energy savings, residents have been willing to "go solar." Michael Deery, spokesman for Hempstead, New York, which is converting to solar energy said to the *New York Times* in 2008, "Our first and foremost goal is to reduce our

SOLAR POWER	
ADVANTAGES	**DISADVANTAGES**
• sunlight is free • quick to install • easy to add on to the system • no pollution from energy production • quiet • little disturbance of land • photovoltaic cells last for several decades	• high costs at present • need access to the Sun about 60 percent of time • needs energy storage system • may need energy backup system • some homeowners do not like solar panels' appearance • takes 40–50 years for energy savings to make up initial cost • manufacturing produces hazardous silicon wastes

carbon footprint and keep our planet clean." Hempstead may be unique in its choice of environment over money. Many other towns would likely not make the same choice, so solar energy joins all other forms of energy in its need to be cost effective.

Is solar energy the answer to weaning the world off fossil fuels? Solar power's detractors cite the large area of land that huge solar panel arrays would require to provide enough energy for the U.S. demand. Arrays can be installed in three different ways: (1) long lines of concave arranged solar panels, called troughs, about 25 feet (7.6 m) off the ground and covering more than 100 acres (40 ha); (2) a large concave dish of solar panels that uses less land; or (3) photovoltaic arrangements that concentrate the total energy output. In 2001, the energy expert Nathan Lewis of the California Institute of Technology warned that the greatest limitation to solar power might be the capacity of the land to contain huge arrays

Solar Power Plants

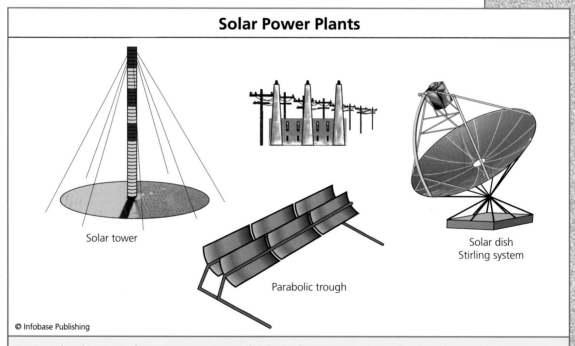

Solar tower

Parabolic trough

Solar dish
Stirling system

© Infobase Publishing

Large-scale solar power plants can use various technologies for increasing their efficiency of converting solar energy into electrical energy. A solar tower uses sunlight-heated air to form an updraft that runs the plant's turbines; a parabolic trough collects direct sunlight and reflected sunlight; a solar dish with a Stirling engine uses a solar concentrator to maximize power. The cost of solar energy has declined in the past 20 years.

of solar collectors. He calculated that if all the solar panels needed to provide energy for the United States were laid flat on the landscape, the panels would cover 66,750 square miles (172,882 km^2) or about the size of the state of Washington.

Arizona owns the world's largest solar power plant, which is in construction about 70 miles (113 km) southwest of Phoenix. The plant, called Solana and planned for opening in 2011, will produce 280 MW of electricity to power about 70,000 homes. To achieve these high levels of production, Solana will use concentrating solar technology that increases the total energy production by supplementing the solar panels with *solar concentrators*. Concentrators use an internal lens to make the Sun's rays less diffuse and more focused. As a result, a power company can put more photovoltaic cells into each solar collector. (Makers of home solar panels may soon use concentrators to increase the efficiency of photovoltaic cells so that solar panel size can decrease, as will costs for homeowners wishing to install solar panels.)

Two emerging solar technologies are solar films, discussed in the sidebar on page 108, and solar satellites. The National Security Space Office (NSSO) located near Washington, D.C., has stated that satellite solar power may be the best way to collect energy from the Sun. In 2007, the NSSO issued a report titled "Space Based Solar Power as an Opportunity for Strategic Security" to introduce the idea of space satellites collecting sunlight and then beaming the energy to Earth. Satellite-mounted photovoltaic cells would capture photons, and then a device would convert the current produced in the solar cells to radio waves or infrared light. The NSSO plans for the satellite to beam the waves to a receiving antenna on Earth connected to an electrical generating utility. The technology depends on expensive lightweight solar panels, the satellite to hold them, a launch vehicle, and transmission and collection instruments. Despite these obstacles, the NSSO values the idea because it could help free the United States from dependence on imported fuels.

The NSSO explained the reasoning behind solar satellites: "Our Sun is the largest known energy resource in the solar system. In the vicinity of Earth, every square meter of space receives 1.366 kilowatts (1,366 watts) of solar radiation, but by the time it reaches the ground, it has been reduced by atmospheric absorption and scattering; weather; and summer, winter, and day-night cycles to less than an average of 250 watts per square

Solar panel technology has been the subject of tremendous interest by researchers and by entrepreneurs seeking improvement or a next-generation solar collector. Though solar panels similar to these in Spain are the most popular device for making buildings sustainable, new solar collector technologies on the way include solar films, ultrathin films based on nanotechnology, solar concentrators, and solar windows. *(Fernando Tomás)*

meter." (Appendix E provides an explanation of common energy units.) The NSSO plans for its space-based solar technology to provide continuous and predictable solar power to Earth by avoiding these energy losses. As for safety, a solar satellite's beams would be of fairly low power on a par with energy emitted from a microwave oven's door. Safety measures might also include no-fly zones in the beam's vicinity and off-limits areas near the receiving antenna.

Solar power technologies have been followed by the media because of the interest solar power has created in the public and in scientific circles. Solar power should be pursued simply because the Earth receives more energy from the Sun in one hour than it uses in one year. Of the 382.7 trillion *terawatts* (TW) of energy emitted by the Sun in all directions, 120,000 TW reach the Earth's surface. Even with losses of solar energy to the universe, this represents an enormous amount of energy.

Some new solar technologies may fall behind due to technical challenges or high cost, but solar power has an advantage over other renewable energy technologies: Solar power has received a great deal of support from the public. The following table describes new solar technologies that may become commercially feasible in the future. Each of the technologies described in the table also has the potential of incorporating solar concentrators.

SOLAR POWER TECHNOLOGIES		
TECHNOLOGY	**DESCRIPTION**	**ADVANTAGE**
parabolic troughs	collection panels arrayed in long straight arrangements	capable of producing large amounts of solar power
solar dish–Stirling technology	concave-shaped solar collector that contains a concentrator and is connected directly to a generator	large energy output
solar tower	collection panels arranged around a tall cylindrical tower heat the air beneath them, creating an updraft into the tower, which provides energy to run attached turbines	potential to multiply the energy output of the solar panels alone; requires no energy input
space-based solar power	satellite-mounted collector panels receive solar radiation, and the satellite beams it to Earth as radio waves or infrared light	collects the maximum amount of solar energy heading for Earth without losses to the atmosphere
thin films	solar cells measuring hundreds of times thinner than solar panels collect solar energy on a variety of surfaces	space-saving and adaptable to more places than panels

SOLAR FILMS

Standard solar panels contain crystalline silicon that must be of a minimum thickness to generate an electrical current. These solar panels take up space atop buildings or on land. Though solar power is rapidly gaining ground in total energy production worldwide, some experts in the field believe the technology may soon yield to new, thinner solar collectors called solar films. Peter Harrop, chairman of the London research firm IDTechEx, told *Time* magazine in 2008, "Crystalline silicon has had its day. These new technologies [films] will be taking over." Harrop's optimism comes from the fact that thin, flexible solar films can be constructed to roll onto a surface similar to wallpaper and replace bulky solar panels with a lower profile appearance.

Thin solar film contains the following four layers: (1) a transparent conducting material exposed to sunlight; (2) a buffer layer; (3) a layer of chemicals such as copper, cadmium, indium, gallium, and diselenide, which produces an electric current from the sunlight; and (4) an underlying contact layer. (Cadmium is a toxic metal that requires safe disposal.) The interface between the buffer layer and the contact layer generates the electrical current. Thin solar film producers are now making films containing these four layers that measure no more than 100 nanometers thick, or about one-thousandth the thickness of a human hair.

Solar film manufacturers have been clamoring to enter this fast-growing segment of the solar power market,and they hope to soon see solar films rolled out onto roofs, walls, and windows.

Engineers at the Massachusetts Institute of Technology (MIT) have been working to combine thin solar films with solar concentrators for the purpose of developing a new technology in windows. The MIT electrical engineer Marc A. Baldo explained in 2008 in the institute's newsletter, "Light is collected over a large area (like a window) and gathered, or concentrated, at the edges." Baldo said the concentrator increases the electrical power from each solar cell "by a factor of over 40." The solar concentrators may have two practical uses that could enter the solar market in a short time. First, concentrators may be set along the edges of a flat glass window panel to generate electricity for indoor use. Second, concentrators may be installed to help boost energy output from traditional solar panels.

Despite improvements that solar power still requires, this clean energy remains one of the most attractive choices for the future for both homeowners and communities.

Thin solar films are easier and cheaper to manufacture than traditional silicon cells, but scaling up to mass production has been difficult. The oil company British Petroleum tinkered with solar films for years before dropping the project in 2002 because of the difficulties of large-scale production. Later, however, solar companies such as First Solar in Arizona with operations in Germany have continued thin film research. First Solar has developed a film using cadmium telluride as an efficient semiconductor layer. The solar film produces equivalent amounts of energy as traditional silicon solar cells but uses only 1 to 2 percent of the raw materials.

By 2008, First Solar completed building North America's largest thin solar film power plant near Boulder City, Nevada. Michael W. Allman, president of the power plant's operator Sempra Generation, said, "This is a significant step in the development and deployment of renewable solar power. The size and scope of this new solar generation facility clearly demonstrates that we can build projects on a scale that helps utilities meet their renewable energy goals." The solar film plant generates 10 MW of power, which could power close to 3,000 homes.

Solar film currently produces less power than large plants fitted with traditional solar panels, but the solar film industry now has at least one company among Fortune 500's fastest growing companies and the best profit growth among all energy companies.

HYDROPOWER AND GEOTHERMAL ENERGY

Hydropower and geothermal energy both make use of energy that is stored in different forms of water. Hydropower uses liquid water; geothermal energy comes from underground sources of heated water or other heat sources in the Earth's crust. Both hydropower and geothermal energy behave as renewable energies because the Earth regenerates water in its water cycle.

Hydropower, also called hydroelectric power, uses the massive amount of energy that exists in large volumes of flowing water. The main way to capture hydropower has been the construction of large dams across rivers to cause the water approaching the dam to build up and form a reservoir. Reservoir water flowing through the dam's pipes, called penstocks, turns turbines, which power generators that produce electricity. Transmission lines carry the electricity to near or distant communities. In the United States, hydropower supplies more than 70 percent of all renewable energy production.

Hydropower accounts for about 25 percent of the world's energy generation, but it accounts for only 7 percent of all energy production in the United States. The West Coast depends on hydropower to a greater extent than the rest of the nation; more than half of all hydropower produced in the United States is produced in Washington, Oregon, California, and Montana. The DOE has calculated that hydroelectric dams in the United States have the capacity to supply electricity to 28 million households, an amount of energy that equals almost 500 million barrels of oil.

Why does hydropower seem to receive less interest than solar power and other renewable energies? Though dams produce clean energy and little pollution, they create other troubles for the environment. U.S. salmon and trout populations have been severely reduced since at least the 1980s, and many environmentalists believe dams have played a big role by preventing the fish from swimming to upstream spawning grounds. Ecologists have experimented with *fish ladders* to provide water routes for fish to bypass dams during upstream migrations. The environmental group Save Our Wild Salmon has argued for removal of many dams to restore fish populations: "... there is no doubt that restoring critical freshwater habitats will increase survivals."

The hydropower industry and environmentalists have differed on other aspects of the environment as well, namely habitat destruction. Construction of a new dam permanently alters riparian ecosystems because dams flood upstream land and alter the natural downstream flow. The following table provides the advantages and disadvantages of hydropower.

Three different types of underground heated water contribute to geothermal energy. Each of these types occurs in the Earth trapped between rock formations, in cracks in the rock, or within porous rock. The three types of geothermal energy are the following: (1) wet steam consisting of hot water droplets and vapor; (2) dry steam containing only water vapor and no droplets; and (3) hot water. Three different types of power plants also exist to convert the heat energy from geothermal sources to electricity:

- Dry steam plants pump steam directly from the underground source to the plant's turbines.

- Flash steam plants extract hot water, convert it to steam, and use the steam to drive the turbines while letting the cooled water flow back underground.

- Binary power plants transfer the water or steam heat to another liquid, which drives the turbines.

Homes can use geothermal energy from water by installing either a heat pump or a geothermal exchanger. A heat pump collects heat from underground sources—sometimes by using a drilled well—in the winter and stores excess summer heat underground. A geothermal exchanger contains a similar system in which buried pipes direct heated water from

HYDROPOWER AND GEOTHERMAL ENERGY	
ADVANTAGES	**DISADVANTAGES**
hydropower	
• large energy output • efficient conversion of kinetic energy to electrical power • low-cost electricity for customers • provides steady source of irrigation water and downstream river habitat • lasts many years with proper maintenance	• high construction costs • flooding to upstream environment displaces wildlife and people • alters natural downstream waterways • interferes with salmon reproduction • halts natural flow of nutrients to downstream habitats • danger of collapse
geothermal energy	
• steady, free energy source • low land disturbance • high efficiency of energy conversion • minimal construction other than the power plant • low pollution	• limited number of sites • sometimes produces little power per well • difficult to store or modulate • emissions, odors, and noise

Pollutant-free geothermal emissions originate from shallow ground to several miles deep in the earth. Geothermal sources give a direct supply of hot water or heat or steam for large-scale electricity. The Nesjavellir geothermal power plant in Iceland began operations in 1990 and produces hot water and electricity for its customers. Residents of Iceland have used the island's rich supply of hot springs for washing, bathing, and heating since the first settlers arrived.

the underground source in winter and carry warm water from the house to the ground in summer.

Geothermal energy also offers a type of energy production called hot rock technology. This type of geothermal energy comes from the following three non-water sources: (1) the Earth's molten rock called magma; (2) hot dry rock, which is rock heated by a magma layer below it; or (3) warm-rock reservoirs, which contain rock heated to above normal temperatures by nearby hot zones containing either magma or steam.

Companies that tap geothermal energy do so by drilling wells into the heat-containing formations called crystalline rock as deep as 16,500 feet (3 miles or 5 km) into the Earth's crust. The power plant pumps water down one of the wells, called an injection well, which forces heated water upward through production wells. An aboveground facility captures heat energy from the water that reaches the surface at about 390°F (199°C). As the water cools, it flows back underground in a separate set of wells. In a sense, a geothermal power plant runs its own miniature water cycle for producing electricity.

Geothermal energy production may be decades away from rivaling nonrenewable energies, but the United States and other countries have geothermal research projects underway. The U.S. Department of the Interior has embarked on a program in conjunction with the Bureau of Land Management (BLM) and the U.S. Forest Service (FS) for building geothermal energy plants. The BLM or the FS will lease different portions of western land that they manage, other than national parks, for power companies to build geothermal facilities. Dirk Kempthorne, former secretary of the interior, said in December 2008, at the time of the program's approval, "Geothermal energy will play a key role in powering America's energy future. All but 10 percent of our geothermal resources are found on federal lands and facilitating their leasing and development is crucial to supplying the secure, clean energy American homes and businesses need." The federal agencies involved in the project expect to generate 5,500 MW of energy from 12 western states including Alaska by 2015 and expand to more than 12,000 MW by 2025.

The Earth's major geothermal sources lie along a line rimming the entire Pacific Ocean across the Northern and Southern Hemispheres. This line, which is called the Ring of Fire, represents major *tectonic plate* boundaries where earthquake and volcanic activity are the highest. Forty-six countries on the Pacific Ocean now use the Ring of Fire for a portion of their energy needs. In the United States, most geothermal energy use occurs in California where 2,500 MW from geothermal sources provide electricity for 6 million people.

Geothermal energy has not grown quickly due mainly to the limited available heat sources on the globe. Advances continue in this area, however, for the purpose of taking advantage of this free and almost limitless energy source. The following table describes three emerging geothermal technologies.

NUCLEAR ENERGY

Nuclear energy may be thought of as a nonrenewable energy source because it uses uranium, which is a nonrenewable element found in the

EMERGING GEOTHERMAL TECHNOLOGIES	
TECHNOLOGY	**DESCRIPTION**
enhanced geothermal system	improved methods in hot rock technology for capturing non-water heat
geopressured-geothermal systems	extraction of natural gas in conjunction with geothermal energy
hydrocarbon-geothermal coproduction	capturing energy from the heated fluids that flow through oil and natural gas reserves

Earth's crust. Scientists can also argue that nuclear energy belongs to the renewable category because a nuclear reaction is a self-sustaining process. Regardless of how a person chooses to view nuclear power, this form of energy has weathered controversies that continue today.

Nuclear reactors are facilities that produce electricity from nuclear fission reactions with uranium 235 and plutonium 239. The reactor produces an efficient supply of power, but it also creates radioactive waste. Radioactive waste poses a serious health threat to living things, and the threat lingers because these materials can remain hazardous for centuries. Finding a safe storage site for radioactive waste has been one of the nuclear energy industry's main concerns. Properly run and maintained nuclear power plants make clean energy without polluting the air, but nuclear energy's disadvantages can be quite serious compared with its advantages as the following table describes.

Opponents of nuclear power cite the three following concerns that they feel override any advantages of nuclear power: (1) the chance of catastrophic accident that contaminates the environment with radioactivity and kills people and wildlife; (2) an accumulation of radioactive wastes that do not degrade to safe levels for thousands of years; and (3) the opportunity for terrorist attack on nuclear power plants or on waste transports. These issues have sustained a long-lasting argument over the merits of nuclear power that is one of the most contentious in environmental science.

The international environmental action group Greenpeace has stated its unequivocal opposition to the risks nuclear power poses to the environment and humanity. "Despite what the nuclear industry tells us," Greenpeace has said, "building enough nuclear power stations to make a meaningful reduction in greenhouse gas emissions would cost trillions of dollars, create tens of thousands of tons of lethal high-level radioactive waste, contribute to further proliferation of nuclear weapons materials, and result in a Chernobyl-scale accident once every decade." The Chernobyl accident was a 1986 explosion in a nuclear reactor in Chernobyl, Ukraine, that produced a radioactive cloud that entered the atmosphere and circled the globe, killed hundreds of people immediately, and cost the lives of an estimated 15,000 people in the following years. Chernobyl remains the world's worst nuclear disaster.

Nuclear power accounts for about 20 percent of the energy used in the United States, produced by about 65 nuclear power plants containing a total of 104 nuclear reactors. Nuclear power offers some undeniable benefits in the quest to reduce fossil fuel use and cut greenhouse gas emissions. A single reactor can generate enough electricity in a year to supply 740,000 households according to the Nuclear Energy Institute (NEI). To do the same using fossil fuels, these households would need 13.7 million barrels of oil, 66 billion cubic feet (1.9 billion m³) of natural gas, or 3.4 million tons (3.1 million metric tons) of coal. In addition to these statistics that

NUCLEAR POWER	
ADVANTAGES	**DISADVANTAGES**
• steady energy supply • uranium resources remain plentiful • low emissions • conserves fossil fuels • established technology • power plants do not require much land • few plants can generate large quantities of electricity	• low energy yield for the costs of operating • potential dangerous accidents • long-term storage of wastes, called spent fuel • parts and equipment contain low levels of radioactivity • requires diligent security • hot process water released into environment injures and kills aquatic life

make a good case for nuclear power, the NEI has pointed out, "Nuclear power plants aid compliance with the Clean Air Act of 1970, which set standards to improve the nation's air quality. Because they generate heat from fission rather than burning fuel, they produce no greenhouse gases or emissions associated with acid rain or urban smog. Using more nuclear energy gives states additional flexibility in complying with clean-air requirements." This positive viewpoint has not calmed the fears of nuclear power's staunchest opponents.

Can nuclear power survive its history and its critics? The plants that continue supplying a significant amount of U.S. energy consumption are beginning to show wear, and many plants will need to be replaced in the next decade. By 2012, about 230 nuclear reactors worldwide and 20 in the

The Cattenom Nuclear Power Station in Lorraine, France. Western Europe leads the world in nuclear power with about 130 nuclear power plants. The number of nuclear power plants increased worldwide between 1970, when the nuclear power industry began, and the mid-1980s. U.S. residents worried about nuclear power after a 1979 accident at the Three Mile Island plant in Pennsylvania. A deadly accident at the nuclear plant in Chernobyl, Ukraine, in 1986 stopped further growth in the industry. Serious flaws in safety and management have overshadowed nuclear power's generally good record, clean energy, and inexpensive electricity. *(Stefan Kühn)*

United States are due for retirement. An aging reactor increases the risk of failure and accidents. Despite the promise of nuclear energy of 50 years ago when the industry began, nuclear power has not made many advances, and it sits in a quagmire of debate involving the public, government, and industry.

Radioactive waste causes concern because of its danger to human health and the very long time required for certain radioactive elements to decay to safe levels. The United States has built a permanent underground storage for nuclear wastes at Yucca Mountain in a remote part of Nevada, but opposition to tons of waste transports traveling by railroad to the site and the overall safety of the facility have delayed Yucca Mountain's opening. Other problems within the nuclear industry since its inception are that nuclear power plant construction costs have exceeded their budgets; the plants have high operating costs; and there has been poor management in safety programs.

Nuclear power's future as an alternative to fossil fuel energy continues to be an unanswered question. The nuclear industry has done a poor job of teaching the public about its technology and new mechanisms for assuring safety. Environmental groups opposed to nuclear power filled the void, and today a significant number of people do not want this form of energy. Nuclear power presents a complex problem. At some point in the near future, scientists, the public, and government leaders must weigh the costs of potential dangers from nuclear power against the real and immediate dangers of climate change.

DIRECT CARBON CONVERSION

Direct carbon conversion is any chemical process that changes one form of carbon into another form with a concurrent production of energy, usually as electricity. Fuel cells have been built on the principle of direct carbon conversion into energy by generating a flow of electrons that convey an electrical current. Innovations in direct carbon conversion have now been proposed that would convert much larger amounts of atmospheric carbon into a usable form.

Fossil fuels, biomass, synthetic fuels, and biodiesel all work as fuels because they contain carbon compounds that release energy during combustion. Energy held within chemical bonds between carbon and other elements, usually hydrogen, serve as the energy-storage form in these fuels. When considering all of the combustion engines in use today and the fact that all living things cannot exist without carbon compounds, it seems as if carbon chemistry truly powers the planet.

Reliance on carbon fuels has caused an accumulation of carbon-containing by-products in the atmosphere in the form of CO_2 and methane. Most people know that these greenhouse gases cause global warming by holding excess heat in the Earth's atmosphere. Less understood is the time that the gases stay in the atmosphere. The *Time* magazine reporter Robert Kunzig warned in 2008, "Once we stop burning fossil fuels, it could take as long as 100,000 years for the CO_2 we've been pouring into the atmosphere to be gone." The Earth's plants, water, and soil soak up a considerable quantity of carbon, but carbon emissions outpace carbon consumption. Photosynthetic organisms absorb CO_2. Some of the carbon also settles in soil or in sediments under the ocean and begins a slow inexorable return to fossil fuel. But Kunzig warned that scientists have found that the ocean and land do not soak up as much CO_2 as they once did, perhaps because humanity's carbon emissions have begun to overload the Earth's natural carbon cycling.

Most atmospheric CO_2 returns to the Earth by absorption into ocean *phytoplankton*, tiny plant organisms that serve as food for millions of other organisms. Phytoplankton levels have decreased in parts of the world's oceans, due to pollution, climate change, and other factors in damaged ocean ecosystems. Some scientists have wondered if technology can find a way to restore the ocean's absorptive ability or even increase it to control atmospheric carbon. The San Francisco company Climos has undertaken a plan to add nutrients to ocean waters to reinvigorate phytoplankton. In this method called iron-seeding, ships will pour iron-rich mixtures into the sea—about 20 pounds of iron per square mile ($3.5 kg/km^2$). The chief science officer of Climos Margaret Leinen told *Time,* "We're not thinking of this as solving the problem. We're looking at this as one of a whole portfolio of techniques." This ambitious plan has yet to be proven as a way to reduce CO_2 in the atmosphere, but climate experts have learned to welcome any innovation to slow global warming.

Other scientists have investigated similar ideas for pulling carbon out of the atmosphere and converting it back into useful fuels. A Harvard graduate student Kurt House has developed a scheme for changing the ocean's chemistry so that it can again absorb very large amounts of CO_2. House's plan involves the following steps:

1. Pump seawater into facilities that split the salt (sodium chloride) into positively charged sodium and negatively charged chloride molecules.

2. Remove the chloride, which would turn the water more basic.

3. Return acid-depleted water to the ocean.

4. The ocean acts to regain its acid-base balance by absorbing more CO_2 from the air.

Allen Wright of Global Research Technologies in Arizona has proposed a third approach in carbon conversion. Wright and physicist Klaus Lackner of Columbia University have built scrubbers to remove CO_2 directly from the air. Their prototype scrubbers each contain about 30 plastic sheets measuring about 9 feet (2.7 m) high. As air moves through the sheets in the scrubber, the CO_2 sticks to the specially formulated plastic. The scientists envision much larger scrubbers and sheets distributed throughout the continents to remove carbon emissions from the atmosphere. Wright remarked in 2008 to the reporter Robert Kunzig, "If we built one [a scrubber] the size of the Great Wall of China, and removed 100 percent of the CO_2 that went through it, it would capture half of all the emissions in the world." Like House and the scientists at Climos, scrubber technology seeks to take on the problem of global warming on an extremely large scale.

The examples of carbon conversion described here have plausibility in laboratory experiments, but no one has implemented them on a grand scale to truly affect climate change. Changing the planet's ocean chemistry represents a monumental job, and the impact of adding large quantities of iron or altered seawater on ecosystems is unknown. Some of the methods also produce large amounts of material that must be managed; House's technique of turning seawater more basic results in large amounts of acid on land that require disposal. For the present, no one has proposed a good solution for managing the excess acid.

But what if these ideas work? Wright has suggested that the CO_2 exiting his scrubbers can be combined with hydrogen to make a new batch of hydrocarbon fuels for cars. Though cars would release more emissions, the scrubbers would simply remove the emissions again and again to create a sustainable carbon loop.

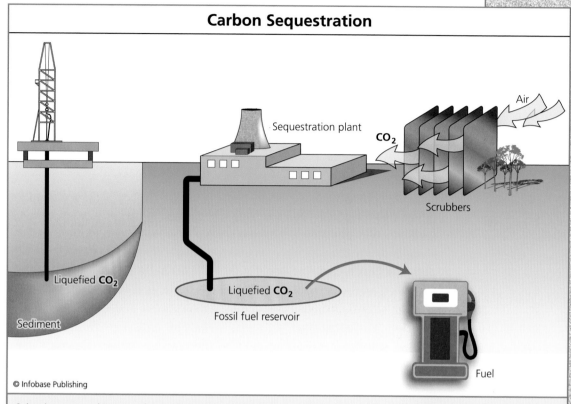

Carbon Sequestration

Sequestration plant

CO$_2$

Air

Scrubbers

Liquefied CO$_2$

Liquefied CO$_2$

Fossil fuel reservoir

Sediment

Fuel

© Infobase Publishing

Scientists are working on a far-reaching plan called carbon sequestration in which devices capture CO$_2$ gas, which can be liquefied and injected into ocean sediments. Cold offshore temperatures will keep the CO$_2$ in liquid form and prevent gas from escaping. The CO$_2$ disposal step may be easier than developing devices to remove large volumes of CO$_2$ from the atmosphere, such as the scrubbers illustrated here.

These methods may not come about in the near future, but they show that innovative thinkers have not been afraid to tackle the environment on a massive scale. Direct carbon conversion such as the scrubber-to-fuel concept might become one of the next-next-generation technologies in sustainability.

FUEL CELLS

Fuel cells convert chemical energy into electrical energy. Fuel cells have been developing rapidly since the 1990s as an alternative energy source for household electronic products or for vehicles to replace batteries and fossil fuel, respectively. The conversion of chemical energy into electricity can be done by either of two types of fuel cells: chemical or biological. Chemical fuel cells produce electricity by running a chemical reaction. Usually heat supplies the initial energy to get the process underway. Biological fuel cells contain one or more components from nature, and enzymes control their reactions rather than high temperature.

Carbon fuel cells react a carbon compound with oxygen to generate an electron flow. Though this type of fuel cell can use carbon-containing wastes as fuel, it also produces CO_2 as its end product. In a carbon fuel cell, the same chemical reactions occur as in combustion, but the entire process runs more efficiently than combustion and generates more energy per unit of fuel.

Hydrogen fuel cells also create an electron flow for producing electrical energy, but these fuel cells use hydrogen rather than carbon in a reaction with oxygen. The advantage of the hydrogen fuel cell over the carbon fuel cell lies in the fact that it produces water rather than CO_2 when generating electricity.

The British physicist William Grove designed the forerunner of the modern hydrogen fuel cell in 1839 based on the knowledge that an electric current could split water molecules into hydrogen and oxygen molecules. By hypothesizing that the reaction could be made to run in the opposite direction, water could be produced and the resulting electron flow would create an electric current between the anode (a positively charged pole) and the cathode (a negatively charged pole) of the cell. Grove's fuel cell performed the following reactions:

$$\text{anode side: } 2H_2 \rightarrow 4H^+ + 4 \text{ electrons}$$

$$\text{cathode side: } O_2 + 4H^+ + 4 \text{ electrons} \rightarrow 2H_2O$$

$$\text{total hydrogen cell reaction: } 2H_2 + O_2 \rightarrow 2H_2O$$

Fuel cell technology has advanced from small reactors, such as that invented by Grove, to high-voltage generators for two main purposes: transportation and electric power production plants. The intended future of large and small fuel cells is to accomplish the following tasks:

- replace gas turbines in power plants
- replace gasoline engines in vehicles
- replace batteries in computers and electronics

Six types of fuel cells that use different internal chemistry have been proposed as one way to reduce dependence on fossil fuels. The following table presents current large-scale fuel cell technology.

Chemical fuel cells offer the advantage of serving as a power generator without the need for fossil fuels. In many cases, these fuel cells do not emit hazardous emissions. But fuel cell technology has also been held back by high costs, high operation temperatures, and inefficiency caused by impurities in the reaction cell.

Biological fuel cells are another emerging technology in energy generation with an, as yet, unknown future. Biological fuels cells use microbes

TYPES OF FUEL CELLS		
FUEL CELL	MAIN FEATURE OF OPERATION	USES
alkaline	reacts pure hydrogen with pure oxygen	space vehicles
direct methanol	alcohol separates the two electrodes and accommodates a current	cars, buses, appliances
molten carbonate	carbonate held at very high temperature 1,112–1,202°F (600–650°C) separates the two electrodes to create current; slow warm-up time	large power plants
phosphoric acid	phosphoric acid separates the two electrodes and accommodates a current; slow warm-up time	medium power plants
polymer-exchange membrane	synthetic polymer separates the two electrodes and accommodates a current with the aid of a catalyst; runs reaction at moderate temperatures 140–176°F (60–80°C)	cars, buses
solid oxide	solid materials provide a matrix for electron flow; slow warm-up time and high temperatures 1,292–1,832° (700–1,000°C)	all sizes of power plants

and their enzymes to act on fuels such as methanol or hydrogen for producing electricity. Biological systems hold advantages over chemical fuel cells because biological systems require no acids or other potentially harmful chemicals, and they run at room temperature.

Entrepreneurs who seek to develop biological fuel cells that can play a role as a renewable energy source have studied various microbes—algae, bacteria, viruses—to carry out the fuel cell reactions. The final result of these reactions will be either biofuels that can replace fossil fuels or electricity production.

The California company Solazyme has used mainly algae to produce biodiesel. Company cofounder Harrison Dillon said in a 2008 press release, "In this search for solutions, Solazyme has taken a 150 million year process of making oil and condensed it to a matter of days to renewably produce oil that can be converted into fuels that not only address these challenges, but have already been proven to be fully-scalable on a commercial level." Assuming this and similar companies can build large *bioproduction* plants, as Dillon suggests, algae may be an important contributor to alternative energy.

Biologists have also experimented with viruses and bacteria for use in fuel cells. Angela Belcher is a biologist who applies her background in electrical engineering and materials science to develop tiny batteries composed of viruses coated in metals that conduct an electrical charge. Viruses measure no more than a few microns in diameter. (A micron equals 1 millionth of a meter.) If virus batteries can be developed for practical use, they will offer the benefits of being very small and light. The virus-metal component might be developed to act as a semiconductor in electronic devices. In 2008, Belcher described her team's work on such semiconductors at MIT: "We have been working on high specific capacity cathode materials using biological processing and getting very good results. We now have full virus-based battery cathodes as well as anodes. We are also working on materials for solar cells, catalysts, fuel cells and carbon sequestration." Recently, Belcher's team has built rechargeable lithium batteries in which viruses construct the battery's conductive material. Belcher has also made *nanoscale*-size wires made of cobalt oxide manufactured by viruses.

Fuel cell technology contains a variety of approaches toward remaking conventional batteries into biology-based batteries. The variety of applications in which these power cells could possibly work also suggests that fuel cell technology has a bright future.

CONCLUSION

Clean energy technology includes an astounding variety of ideas for non-fossil fuel, nonpolluting power supplies. These technologies currently have an encouraging amount of support from government agencies, leaders, and universities. Not every technology will prove to be realistic due to high costs or difficult technical challenges, but certainly a number of new clean energy technologies are possible and on the horizon.

For the most innovative approaches in alternative energy to succeed, the public must understand the advantages and disadvantages of each technology. Leaders in the renewable energy industry must assure skeptics that they are working to reduce the effects of each disadvantage and to offer the promise of a new way of traveling, switching on a light, or heating homes. The most successful alternative energies will also likely be those that provide a seamless change for consumers. If a person clicks on a computer, that person should not notice any difference in whether the electricity comes from a solar power plant or a hydroelectric dam or even a coal-fired power plant. Low costs, ease of use, and minimal disruption to daily routines offer the best ways to win customers.

With this business plan in hand, the clean energy industry can be expected to make the following breakthroughs in the near future: solar concentrators to make solar power more efficient; a growth in commercial solar power plants; the emergence of commercial wind farms; a practical fuel cell for use in vehicles or electronics; and improvements to solar films. Clean energies, especially wind and solar energy, also require a way to store the energy they collect for when it is needed. Currently, few methods exist for capturing and retaining wind energy for when the skies are calm or for solar energy at nighttime. Many other clean energy technologies represent projects for the next generation of scientists. All of these long-term technologies already have research behind them that scientists will build on: solar power innovations such as solar towers and solar satellites; continued growth of geothermal power; the emergence of wave and tidal power. In the far future, scientists may invent mechanical devices for pulling CO_2 directly out of the air and discover a means to return health to ocean ecosystems.

Clean energies have an exciting and promising future in green technology. People outside the environmental sciences can take comfort in the realization that ideas in clean energy have been developing faster than natural resources disappear.

GREEN BUILDING DESIGN

In the 1970s, a person shopping for a new house would probably have found plenty of attractive homes on landscaped property with up-to-date conveniences. The house's water heater kept a supply of hot water at the ready. Large windows opened to splendid views, and electric lighting bathed each room in light, even on sunny days. An air-conditioning unit and an oil furnace assured that indoor temperatures never deviated from a comfortable range. Of course, it took several minutes of running the tap before hot water reached its destination; heavy winds sometimes whistled through gaps in window and door seals, and storms knocked out the electricity. A garbage hauler picked up trash once a week, and that was a good thing because convenience foods and a variety of household products and electronics produced large amounts of discarded packaging. The packaging, wastepaper, bottles and cans, and kitchen wastes all went out in a single trash can set by the curb.

No one spoke in terms of *green building* in the 1970s. Environmentally aware people had begun to point out that natural resources had become strained, but the average family household did not see a problem. Trees ringed neighborhoods, cars went fast, and every year or so stores offered a new model radio or television. People who were serious about conserving a healthy environment participated in local recycling programs, if one existed in their town. Environmentalism extended little beyond recycling.

Construction in the United States provided a high quality of living compared with other parts of the world, but it also brought waste of all kinds—energy, water, land, raw materials, and reusable materials. By the 1980s, people had come to understand the problem of waste. Hazardous wastes had been accumulating in the environment, and by the 1980s terrible health problems began to surface, often because of chemicals in the environment. Decades of poor waste and resource management caught up with many communities. The green building movement grew out of people's desire to take more care in the use of resources, particularly resources that were impossible to replenish in a human lifetime: forests, metals, minerals, and possibly clean air and clean water.

Green building refers to a way of constructing buildings without making the mistakes earlier generations made because they did not fully realize the effects their actions had on the environment. The green building process can be divided into six main focus areas, all of which contribute to a healthier living environment for people and create minimal disturbance to the natural world. These six processes are (1) energy use, (2) land and water use, (3) materials, (4) construction methods, (5) integration with the community, and (6) indoor environmental quality. Green builders today take all six focus areas into consideration when planning new construction.

Green builders and designers also target the following three objectives in every new project they begin: (1) construction planning and landscaping to minimize a building's effect on ecosystems; (2) maximizing the use of recycled materials and reducing construction wastes; and (3) creating a structure that enhances the environment.

To meet these objectives, green builders break down the six individual construction focus areas into more detailed steps. A decision at each step in the building process allows a builder and a homeowner to choose materials or methods that are least harmful to the environment. It follows therefore that one green building can differ quite a bit from another green building a few doors away. One building may perhaps contain 90 percent recycled building materials, with the remaining materials made from only certified sustainable sources that do not cause an overall detriment to the environment. A second homeowner may be less concerned with building materials, but might emphasize renewable energy sources by selecting the most efficient insulation and windows and adding a *gray water* reuse system.

This chapter outlines the various decisions used by builders and owners in constructing a building that exerts minimum effect on the environment. Every decision made with intent to help the environment gives a benefit, even if the benefit is small. Therefore, people do not have to build a 100 percent renewable energy house that recycles all its wastes. The environment will also receive benefits when a person chooses certified sustainable woods; recycles glass, paper, plastics, and aluminum; and turns off appliances when not in use. This chapter begins with a brief background of green building, then examines specific factors that make a building "green." The main factors covered are energy and heating, cooling and ventilation, insulation, lighting and windows, water conservation, and waste management. The chapter also describes an ideal off-the-grid house and includes a case study on one such house that follows green building principles.

BUILDING GREEN COMES OF AGE

A green building plan starts by conforming to the Three Rs: reduce, reuse, and recycle. Keeping these actions in mind helps building designers lessen the amount of materials used, the wastes they produce, and the impact a new building will have on the ecological footprint. Fortunately, today's recycling industry has made rapid gains so that almost any virgin building material can be replaced with a sustainable material that often has better qualities.

Attention to building materials began in the United States in the 1930s with the introduction of air-conditioning, fluorescent lighting, and structural steel. Architects designed buildings that disconnected inhabitants from the outdoors by installing powerful heating-ventilation-air conditioning (HVAC) systems. People developed the idea that their lives had nothing to do with nature. The construction industry simultaneously began to separate into specialties: design, architecture, construction, and civil engineering. These professionals brought inventive ideas to their work, but a lack of communication led to projects that did not take into account a building's entire purpose or its relationship with the environment.

GreenBuilding.com has explained that not until the 1970s did "a small group of forward-thinking architects, environmentalists, and ecologists inspired by the work of Victor Olgyay (*Design with Climate*), Ralph Knowles (*Form and Stability*), and Rachel Carson (*Silent Spring*) [begin] to question the advisability of building in this manner." The American Institute of Architects (AIA) responded to a U.S. energy crisis in 1973 that had started with rising oil prices from foreign suppliers. The AIA built a decades-long program of seeking the best solutions in solar energy, waste

The Opus One Winery in California's Napa Valley was not conceived as a green building in the early 1990s. However, the building captures many elements of today's green designs: partially underground for insulation and cooling, use of daylighting, natural materials, and water conservation. *(Chuck Szmurlo)*

reduction, water conservation, and sustainable materials. By 1993, the AIA had chosen sustainability as its theme for the International Union of Architects/AIA World Congress of Architects meeting. This event has often been cited as the turning point where the green building movement became an industry.

From the 1990s onward, private and government institutions have published several design guides for green building. Professional associations have launched international design competitions throughout the decade. In 1999, President Bill Clinton signed Executive Order 12852 that established a Council on Sustainable Development. This council issued a report naming 140 actions that could be taken by U.S. residents for environmental improvements, and many of these related to building sustainably.

Today, hundreds of builders, designers, and architects and dozens of professional associations offer tips and training on green building. All of this ready information has perhaps made some people lose focus on the concept of green building in a rush to join the green trend. The *San Francisco Chronicle* magazine writer Jane Powell offered this opinion in 2007: "Building or remodeling uses up resources, even if those resources are recycled or salvaged. The greenest thing you can do is continue the life of an existing building, whose resources have already been extracted … There is an all-too-common practice of demolishing a small existing building in order to throw up … a larger 'green' building, as though the small building had volunteered to be the … sacrifice on the altar of 'smart growth.'" Powell made a valid point. Indiscriminate green choices do not help the environment any more than does replacing a two-year-old sport utility vehicle with brand-new hybrid car. Green building has furthermore become trendy, and homeowners do not always understand the decisions that matter most to the environment.

Powell pointed out that in 1970 the average house measured 1,500 square feet (139 m²), but by 2007 it had reached 2,200 square feet (204 m²). Some well-to-do homeowners have built 6,000-square-foot (557 m²) behemoths, or larger, that they tout as green buildings. These homeowners may believe they are helping the environment by building big structures out of sustainable materials, but the energy costs of construction and maintaining these homes contribute to an overdraft on humanity's ecological footprint.

The following table summarizes some actions that the green building movement has made commonplace in today's construction projects. It also lists problems that continue to arise in construction, indoor activities, or locale.

Green building is a correct and essential choice in sustainability. But the decisions that go into building green must be weighed carefully in order to meet the desired purpose. The sidebar on page 131, "Leadership in Energy and Environmental Design (LEED)," explains how people receive guidance on making good green decisions in buildings and avoid mistakes that spoil the environment.

CONTROLLING ENERGY AND HEAT FLOWS

A building's energy and heating system consists of the energy-generating unit, an energy and heat storage, and the distribution lines. Many items increase the efficiency of how energy is captured, stored, and distributed as heat or as electricity. Some examples of products that make an overall energy-heating system more efficient are smart appliances that regulate power during peak usage times, smart electronics that shut off or decrease their power usage when they are idle, and instant water heaters that save energy and conserve water.

GREEN BUILDING TRENDS AND PROBLEMS TO BE SOLVED	
ACCEPTED OR BECOMING ACCEPTED	**CONTINUING PROBLEMS THAT NEED A SOLUTION**
in the home	
• long-life fluorescent lighting • effective insulation • insulating windows • recycled construction materials salvaged fixtures, metals, and countertops • waste reduction • solar energy	• electricity from fossil fuel–powered power plants • electricity waste • always-on electronics • oversized houses • high per capita energy consumption • excess packaged products
in the community	
• hybrid vehicles • greater use of mass transit spare-the-air days (reduced use of personal vehicles with increased use of bicycling or mass transit)	• limited mass transit • long commutes • dependence on gasoline vehicles • inefficient recycling programs

Inhabitants of a building can also control energy use by their behavior. The main actions that help conserve energy are the following, but energy companies frequently add new energy-saving tips to this list.

- shutting off lights and electronics when not in use
- plugging units such as computer systems and entertainment systems into a dedicated power strip for turning on and off the entire system
- setting temperature to a range of 65 to 68°F (18–20°C)
- shutting any heating vents that are not being used

(continues on page 134)

LEADERSHIP IN ENERGY AND ENVIRONMENTAL DESIGN (LEED)

LEED is a program offered by the U.S. Green Building Council located in Washington, D.C., for certifying sustainably built houses, schools, public buildings, and business structures. The LEED program contains a Green Building Rating System and a related certification program that identifies how successfully new buildings have adhered to the principles of sustainability. Today, the following professions volunteer to follow the LEED guidelines for certification of new or remodeled buildings: architecture, design, interior design, landscaping, engineering, and construction. The real estate and lending professions also pay close attention to buildings that have attained LEED certification because certification makes the buildings more desirable to communities. Many local governments and universities now require LEED certification of new construction.

LEED contains individual rating systems for the following nine different types of construction: new buildings, existing buildings, commercial building interiors, building cores and shells, retail projects, schools, health care facilities, homes, and neighborhood developments. (Even a parking garage in Santa Monica, California, has earned a LEED certification.) A new house, for example, could receive LEED points based on decisions in the areas of design, location, water efficiency,

The renovated building at 330 Hudson Street in New York City is a LEED silver-certified mixed-use building, containing offices, shops, and a hotel. It is one of the few LEED-certified hotels in the United States. (*World Architecture News*)

(continues)

(continued)

energy use, emissions, materials, and indoor air quality. Remodeled buildings may earn LEED certification similar to new building certification.

The U.S. Green Building Council awards points in a variety of categories, so every new building has its own unique way of attaining a certification. The following general point categories each have many specific areas for achieving sustainability:

- innovation and design process—designs to reduce construction and energy waste with lowered costs and compact structures
- location and linkages—appropriate acreage for size of building and access to public transportation
- sustainable sites—landscaping to maximize heating and cooling efficiency, surface water management, and nontoxic pest control
- water efficiency—water reuse, no leaks, high-efficiency irrigation, on-demand water heaters
- energy and atmosphere—reduction of carbon dioxide-releasing systems, good indoor air quality
- materials and resources—recycling, minimized packaging, recycled building materials, biodegradable products
- indoor environmental quality—ventilation, venting, air filtration, low-emission paints and carpeting, maximum use of daylight, radon protection
- awareness and education—promotion of the certification steps taken by designers, builders, and homeowners

The following table summarizes the certification levels for houses.
In order for a building to receive LEED certification, a builder must take the following steps:

1. Join the LEED program.
2. Build the structure to the stated goals.
3. Receive inspection from an official LEED rating grader.
4. Sign forms attesting to accountability for maintaining LEED performance.
5. Receive final certification from the U.S. Green Building Council.

LEED CERTIFICATION LEVELS FOR HOMES		
CERTIFICATION LEVEL	NUMBER OF LEED POINTS REQUIRED	MAIN FEATURES
certified	45–59	• substantial energy savings • minimized construction waste • good insulation • double-paned windows • elimination of water, heat, and electricity waste
silver	60–74	• energy savings in all systems • use of recycled building materials for most of the structure • water efficiency • heating and cooling efficiency
gold	75–89	• majority of furnishings and building materials salvaged, refurbished, or reused • high-efficiency or gray water reuse system • excellent indoor air quality
platinum	90–136	• maximum efficiency in energy and resource use • superior indoor comfort and lighting • dramatic reduction in carbon dioxide emissions • zero or near-zero waste • most or all systems disconnected from the power grid
total available points 136		

Source: U.S. Green Building Council

(continues)

Certification need not be confined to urban areas or affluent neighbor-hoods. LEED principles apply to rural homes, urban or suburban houses, inner-city dwellings, single- or multifamily housing, and rental properties. Each year the Green Building Council updates its requirements for LEED certification in all of its point categories in order to stay current with new sustainable technologies.

The percentage of new projects applying for LEED certification has topped 50 percent since the late 1990s; since 2005, the number of regis-tered and certified projects has increased 250 percent. Most important, the LEED program helps with the design of buildings in which all of a struc-ture's features—heating, lighting, water use, etc.—have been coordinated for maximum efficiency. The LEED program makes the phrase "green build-ing" more than a token; certification notifies everyone that construction has been planned and implemented to reduce the ecological footprint.

(continued from page 130)

- setting clothes washers to warm or cold water washes
- washing only full loads in clothes washers and dishwashers
- air drying washed clothes whenever possible
- when running water to warm it, using the excess water for plants
- using appliances before 9:00 A.M. or after 6:00 P.M. to avoid peak usage times

Green buildings contain innovations that help monitor energy usage and distribute heat in an as efficient manner as possible. Nongreen build-ings have for decades relied on gas or oil furnaces for heat. Green buildings substitute renewable energy sources for gas and oil, mainly by using roof-mounted solar panels. Solar energy, heating, or cooling may be derived by either passive or active means.

Passive methods rely on natural processes, such as sunlight for heating and breezes for cooling and ventilation. Passive heating involves putting large windows on south-facing walls and using heat-absorbing materials for walls and floors.

Active methods use energy sources such as solar and also include devices to store and distribute the energy collected by solar panels. The following two types of active systems can be used in which solar collectors act in combination with heating or cooling processes: hot water collectors that heat the water circulated inside them or air collectors that heat air to be distributed throughout a building by fans.

Wood furnaces and geothermal heat pumps also play a role in green buildings, usually homes rather than larger public buildings. The following table describes the passive and active systems that supply heat energy to households.

Households can better control their HVAC if the house's size fits its use. In other words, a four bedroom house with two bathrooms may suit a family of six people, but the same house wastes space and HVAC energy when occupied by only two people. Most energy distribution and heating systems in green buildings work best when the system is appropriately sized for the size of the structure.

Green building heating systems offer easy maintenance and have a long period of use. Passive systems save more energy and money than active systems, but houses that rely on passive systems require extra planning in design and orientation on the landscape. To make optimal use of passive energy systems, architects and builders incorporate the following features:

- large south-facing windows to maximize capture of sunlight for heating and lighting
- retractable shading to block some sunlight in hot seasons
- proper insulation to support HVAC efficiency
- orientation of house on the site to maximize solar panel efficiency and passive heating

Even large buildings have been successful in the principles that help make houses green. For example, on the campus of Harvard University in Cambridge, Massachusetts, the Blackstone office complex uses a solar thermal system for generating heat. Roof-mounted solar collectors transfer the Sun's heat to tubes containing a continuously flowing antifreeze

GREEN BUILDING HEATING SYSTEMS

SYSTEM	HOW IT WORKS	ENERGY USE
air-source heat pump	draws heat from the outside to the indoors by air-to-air transfer or air-to-water transfer	active
fireplaces and stoves	burning wood or other biomass for space heating	active
forced air furnace	efficient conversion of gas to heat with capability to quickly change a room's temperature and ventilate	active
geothermal heat pump	collects heat from the ground and moves the heated air into the house	active
hydronic radiant heating	hot water is forced through radiators located throughout a house	active
off-the-grid photovoltaic	batteries store energy captured by photovoltaic cells until needed, with no input from the local power grid	active
solar water heater	solar heating of household water rather than oil or gas heating	active
space heaters	small heaters directly connected to solar panels that heat water/air and release heat to a limited space	active
thermal mass	brick, masonry, tile, and concrete absorb heat during the day and slowly release heat indoors at night	passive
windows	double-paned windows and special glazes retain heat	passive

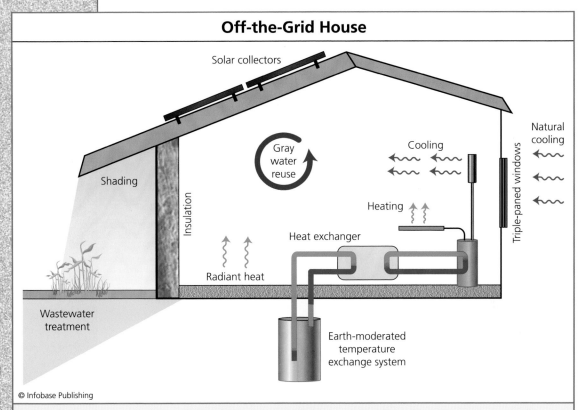

Off-the-Grid House

Solar collectors

Gray water reuse

Cooling

Natural cooling

Shading

Insulation

Triple-paned windows

Heating

Heat exchanger

Radiant heat

Earth-moderated temperature exchange system

Wastewater treatment

© Infobase Publishing

Green building designs offer a variety of energy and natural resource conservation features. A 100 percent off-the-grid house relies entirely on its capacity to generate electricity, heat, and light without drawing from the local power utility. An increasing number of companies sell products that support off-the-grid living and help homeowners recycle gray water, decontaminate hazardous wastes, recycle materials, and manage heating, cooling, and ventilation.

fluid. (Antifreeze protects against temperatures that are too hot or too cold, depending on season.) The fluid circulates to the building's basement where a *heat exchanger* transfers the heat from the antifreeze to the building's hot water system.

The Public Utilities Commission (PUC) in San Francisco has begun planning a 12-story building with additional features. The following green technologies will become part of the new building: roof wind turbines for generating energy; solar panels to supplement the electricity needs; faucet sensors and on-demand water heaters; and recycled gray water. The PUC manager Anthony Irons said in 2007, "I wanted us to design a building completely unconnected to the electrical grid." Power companies and other businesses can set similar examples that show green building is not confined to small dwellings but works in almost any structure.

COOLING AND VENTILATION

Like heating, cooling and ventilation have passive methods to save electricity. Passive cooling employs carefully placed window overhangs for shading, windows with reflective coatings that reduce sunlight, and orientation of a building to take advantage of breezes through open windows.

Cooling a green building follows the same principle as heating a building: Passive methods should do most of the cooling supported by simple rather than complex active systems. Structures that help keep a building cool in hot weather are the following: shading overhangs and shade trees, opened screened windows at night to admit cool air, ceiling fans in place of air conditioners, and good insulation. Residents of green buildings use additional tactics to keep the indoors cool, such as: avoiding the cooling of unoccupied rooms; relocating appliances (freezer, washer, and dryer) to the basement or garage so they do not heat the indoors; minimizing the use of the oven on the hottest days; running appliances such as washers and dryers in the evening; venting the clothes dryer to the outside; and using exhaust fans to vent bathroom or shower humidity to the outside.

Homeowners who use electric air-conditioning can reduce energy consumption by using the system only when needed, keeping the unit well-maintained, and replacing older window or outside units with new energy-efficient models.

Green buildings should contain air-conditioning systems that have one or both of the ratings shown in the following table. A voluntary program for appliance manufacturers called EnergyStar labels air conditioners with a rating that indicates the unit's level of energy efficiency. Consumers should select air conditioners that have been rated at or above the ratings described in the table on page 139. EnergyStar-rated air-conditioning units contain other useful options for conserving electricity: variable speeds, a fan-only option, and a replaceable filter.

Owners of green buildings may decide to use either one of two alternative air conditioners: evaporative coolers or ductless (also called split-system) air conditioners. Evaporative coolers spray a fine water mist into the house that cools the interior as the water evaporates. Ductless air conditioners consist of tubing that circulates cool refrigerant from an outdoor

AIR CONDITIONER (AC) ENERGY-SAVING RATINGS		
ENERGY-EFFICIENCY RATING	CALCULATION	ENERGYSTAR REQUIREMENT
energy efficiency ratio (EER)	cooling output in British thermal unit (Btu) divided by power consumption in *watt-hours* (Wh)	central AC: 11 room AC: 9.4
seasonal energy efficiency ratio (SEER)	seasonal cooling output in Btu divided by seasonal energy input in watt-hours for an average U.S. climate	central AC: 14 room AC: not rated by SEER

unit to individual rooms. Each room contains a small fan that transfers air cooled by the refrigerant into the room's interior.

Ventilation has always been the easiest of tasks: Open a window. But many modern buildings contain sealed windows and rely completely on air-conditioning and heating to regulate indoor temperatures. This method pays a large price in energy consumption. Green buildings therefore contain innovative ventilation systems that minimize energy waste.

Buildings can use four different types of energy-saving ventilation systems. The types are listed here in order of energy demand and cost, from lowest to highest. First, natural ventilation occurs with open windows and through faulty seals around windows and doors. This type of ventilation has a zero energy demand but the building's inhabitants cannot control it other than shutting a window. Second, exhaust-only ventilation removes indoor air to the outside and helps control indoor humidity, but it depends on natural ventilation for the input of fresh air. Third, balanced ventilation includes one-way exhaust fans and one-way intake fans that bring in fresh air. Fourth, central ventilation consists of a system that pulls fresh air into a building at the same rate that it exhausts stale air. Central ventilation in green buildings often includes a heat exchanger that aids in temperature regulation. Many green buildings combine solar power with their HVAC systems to reduce energy use and cost.

INSULATION

Insulation serves a critical role in energy savings by reducing the work that heating and cooling systems must do to regulate an indoor environment. Heat travels through a building in the three following ways:

- conduction—heat transfers directly through materials from molecule to molecule
- convection—heat transfers in air or water
- radiation—heat travels from a hot surface through the air to a cooler surface

Insulation acts to reduce heat loss by conduction. Good insulation blocks the movement of heat from a warm material to a cool material, or vice versa. Therefore, in winter insulation keeps a building's walls and floor from conducting heat to the outdoors, and in summer it does the opposite by preventing heat from coming indoors.

Insulation had for many years consisted of foams made of compounds called chlorofluorocarbons (CFCs) and hydrochlorofluorocarbons (HCFCs). These compounds have been shown to be hazardous to health and to the atmosphere, so green builders now avoid them. Green buildings have many options for materials that insulate well and do not cause health concerns. In some instances, these materials are also recycled from other uses and thus help in waste reduction. The following table lists recommended insulation materials for green buildings. Each material correlates to an R-value that conveys the *thermal resistance* of the material, meaning the material's ability to slow heat transfer. The higher the R-value, the better an insulation material slows heat transfer from one substance to another.

Of the insulation options available today, many come from biological sources. Cellulose is a fibrous molecule in plants that makes up part of newspapers and cardboard. Cellulose insulation can thus be made from recycled newspaper and corrugated cardboard boxes. Biological insulation, such as soybean–based materials, may be of open cell or closed cell variety. This refers to the level of processing done to the shells before the material becomes insulation. Biological insulation should be selected over foam whenever possible because foams produce some greenhouses gases during manufacture. Regardless of the insulation used in a green

GREEN BUILDING INSULATION MATERIALS

Insulation Material	R-value per inch (2.54 cm)	Common Uses	Advantages
batt (synthetic fiber insulation)			
fiberglass	2.9–3.8	walls, floors, attics	easy installation
cotton	3.0–3.7	frame studs, joists, beams	
loose fill			
cellulose, dense pack	3.4–3.6	walls, ceilings, attic floors	good for irregularly shaped areas and hard-to-reach places
fiberglass, dense pack	3.4–4.2	walls, ceilings, attic floors	
mineral wool (rock wool)	2.2–2.9	walls and ceilings that need air sealing	
sprayed insulation			
polyurethane foam	5.6–6.2	walls, attics, floors	seals air as well as insulates and covers hard-to-reach places
polyicynene foam	3.6–4.3	walls, attics, floors	
damp-spray cellulose	2.9–3.4	walls, attics, floors	
spray-in fiberglass	3.7–3.8	walls, attics, floors	
foam board			
expanded polystyrene	3.9–4.2	basement masonry walls and floors	high insulating value per little thickness; covers studs and cavities
extruded polystyrene	5.0	basement masonry walls and floors	
polyisocyanurate	5.6–7.0	exterior walls	
polyurethane	5.6–7.0	exterior walls	
phenolic insulation, closed cell	8.2	exterior walls	
phenolic insulation, open cell	4.4	exterior walls	

Source: David Johnston and Kim Master. *Green Remodeling—Changing the World One Room at a Time* (Gabriola Island, British Columbia, Canada: New Society Publishers, 2004).

building, laws require that all insulation contains fire retardants to reduce the chances of a combustible material burning. Fire retardants may be added to insulation during manufacture or sprayed onto the insulation's surface before installation.

DAYLIGHTING

Daylighting consists of maximum use of natural light indoors to minimize the need for artificial electrical lighting during the day. Green buildings combine a number of strategies to bring more light from outside into all corners of the building. All of the following daylighting techniques require no energy once they have been constructed.

- clerestory windows—horizontal, narrow windows set high in walls to capture low winter sunlight
- light shelves—horizontal, reflective plane near a window that reflects light deeper into a room
- light wells—windowed shafts running vertically along the outside of a building
- reflective surfaces—glazing that reflects light deeper into a room
- solar tubes or pipes—tubes that run from a small, domed skylight on the roof, through an attic, and opening in a room's ceiling to admit light to inner rooms
- skylights—wide, rooftop installations to admit light through the ceiling
- window orientation—east- or west-facing windows to increase natural lighting
- window placement—height of windows to maximize sunlight relative to a building's latitude in the hemisphere

A good daylighting system should minimize electricity use and support heating systems. Green building designers use computer programs to predict the amount and direction of sunlight that a building will receive during each season. With these results, designers can place and orient windows to provide the best daylighting conditions. After they have planned the optimal placement of windows, architects then include other structures such as skylights and clerestory windows to supplement natural sunlight coming into the building.

Energy utility Pacific Gas and Electric (PG&E) has supported a program called the Daylighting Initiative to serve the two following purposes: to encourage better building designs for daylighting and to raise awareness of the benefits of daylighting. In 1999, the Daylighting Initiative supported a study to determine the effects of improved daylight exposure on students in schools. PG&E stated in its report *Daylighting in Schools* that the study "established a positive correlation between higher test scores and the presence of daylight in classrooms." Since that study, others have shown that daylighting leads to improved visibility, mood, and behavior in students during school hours, leading to improved learning.

In addition to schools, other types of buildings have improved daylighting for the purpose of increasing attention, interest, and a general sense of well-being among their patrons. The following types of buildings have explored ideas for better daylighting: museums, retail stores, supermarkets, office buildings, doctors' offices, and athletic clubs.

Green building designers take into consideration the surroundings of the new structure. Green buildings take the best advantage of an area's sunlight patterns, breezes, soil, and climate. The most successful green buildings work in harmony with the environment. *(Lake Attractions)*

Emily Rabin wrote for GreenBiz.com in 2006, "A good daylighting design can save up to 75 percent of the energy used for electric lighting in a building." But daylighting means more than adding a lot of windows to a building. The HVAC engineer Eric Truelove told Rabin, "The biggest problem with daylighting design is we still take a traditional building approach to a daylighting project." Truelove made the point that the best possible daylighting happens only if architects, engineers, and builders work together to understand how light travels inside a building. For example, large windows let in sunlight but can also admit significant heat and glare. As a result, the building's occupants may close the window blinds and resort to electric lighting. For these reasons, daylighting has evolved into an important specialty in green building design.

WINDOWS TECHNOLOGY

Current windows technology plays a part in green building heating, cooling, insulation, ventilation, lighting, and electrical use. Poor windows technology subtracts from the gains made by selecting the right insulation, building materials, and other components of a green building.

Windows and solar energy work in concert to make a sustainable structure because windows have a significant impact on a building's energy conservation. In conventional buildings, energy lost in the form of heat escaping through windows accounts for 25 percent of the entire building's heat loss. This leads to excess energy costs for the building's owner and more energy demand from the power grid.

The efficiency of how well windows contribute to indoor comfort can be helped by three components of the building itself. First, low-conducting window frames made of wood, vinyl, or fiberglass rather than aluminum or steel conduct less heat. Second, windows aligned with materials that possess thermal mass help the window and the material work together in admitting and holding heat. Third, window coverings and retractable overhangs help windows admit heat in the winter and repel heat in the summer.

The technology of window glass has also advanced in the area of energy conservation. The following table summarizes the most promising window technologies used today in green building.

WINDOW TECHNOLOGIES FOR ENERGY CONSERVATION	
TECHNOLOGY	**DESCRIPTION**
glazings	reduces unwanted heat or maximizes heat transfer, depending on type of glaze and orientation of the window
high transmission	usually used in combination with low-emissivity windows to allow maximum sunlight to cross the glass
low-emissivity (low-E)	thin coating or a tint on the inner glass reflects heat back into a room in cold climates; or on the outside glass in hot climates
multiple-paned windows	double or triple panes of glass insulate twice as much as single-paned windows; argon or krypton gas between the panes slows heat transfer
superwindows	thin plastic films between glazed windows of three or more panes

Windows receive ratings called a U-value, which is the inverse of insulation's R-value.

$$\text{U-value} = 1 / \text{R-value}$$

Builders select windows based on the amount of light and heat a building or a room needs. Most green buildings use windows with a U-value close to 0.20. Low-E windows have U-values of about 0.35; superwindows range in U-values from 0.15 to 0.30. Builders and homeowners choose the type of technology based on the sunlight a building receives and its predominant climate. Solar heat gain coefficient (SHGC) offers another rating used by window manufacturers to describe how well a window transmits sunlight. Low SHGC values, on a range between 0 and 1, indicate less sunlight is transmitted, and high SHGC values indicate more sunlight is transmitted.

Solar films for windows differ from the solar films that are beginning to replace solar panels. Window films consist of thin coatings over the glass for the purpose of helping temperature control and for blocking transmission of ultraviolet (UV) light. UV radiation has been linked to certain cancers, especially skin cancer, and it also fades furniture and carpets. The International Window Film Association based in Martinsville, Virginia, lists the following attributes of window films:

- blocking of up to 99 percent of UV radiation
- reduced heat transmission
- decreased glare
- scratch and shatter resistance

New types of windows may someday contain solar cell glass in which the manufacturer imbeds extremely thin solar cells into the glass pane. The windows then become part of the building's total solar energy use. The advent of liquid crystal display (LCD) screens and nanotechnology have made solar cell windows a possibility for the near future. Charles Gay, a manager in the solar film industry, told Reuters in 2007, "The efficiencies [sunlight to energy] are climbing for the thin films. They haven't been around as long [as solar panels] and we're still in the learning phase." Success in the development of solar cell windows will be an important step forward in maximizing the total solar energy that can be captured and used by a green building.

Windows technology works in concert with the construction materials and other factors, such as building orientation, to make a green building reach its energy-efficiency peak. The following sidebar "Case study: Four Horizons House, Australia" describes a famous example of a green building that combines the best selections in materials, lighting, waste management, and locale for reducing energy consumption.

WATER CONSERVATION

A building's clean freshwater source ranks as high in importance as its energy source. Any living organism cannot exist long without water. Green buildings incorporate techniques to conserve on freshwater demand by reusing some water and capturing rainwater.

CASE STUDY: FOUR HORIZONS HOUSE, AUSTRALIA

In 1993, the architecture professor Lindsay Johnston of Australia's University of Newcastle began to design a 2,620 square foot (243 km²) house to resist the quick-spreading bushfires that often threatened rural Australia. In addition to selecting fire-resistant materials for the house, Johnston developed a plan for powering the house without any reliance on the local power grid. To help save energy, the builders oriented the house to make best use of heating, shading, and protection from winds and also to enjoy views in every direction—the four horizons.

In a 2003 interview with *NineMSN* in Australia, Johnston elaborated, "The Four Horizons House and lodges are completely off the grid, so we make our own electricity from photovoltaic panels with a backup generator. We also have solar power radio telephone. Also, we harvest our own rainwater and recycle gray water from the laundry and showers to use for subground irrigation for our vegetable garden. But the rest of the technology is very simple." Four Horizons's additional features consist of a solar-powered water heater, a propane-powered stove and refrigerator, a wood-burning fireplace, lowered walls for cross-ventilation, and a breezeway through the center of the house that helps ventilate and cool the house. The house also relies on a double-roof system for natural cooling in which breezes flow through a space between the lower roof and the larger upper roof.

The energy independence that Johnston built into Four Horizons House has been valuable: The house is in an area of eastern Australia where municipal services are very limited. But rather than equate off-the-grid living with isolation from the media, Johnston installed a solar-powered telecommunications system for phone and Internet connections.

Like most green buildings, Four Horizons is composed of materials that not only provide shelter but serve a purpose in heating, cooling, and lighting. Minimal walls allow light to reach every part of each room. The house's insulation consists of polyurethane and wool. Johnston also included thermal mass materials such as finished concrete floors, concrete block walls, and brick surfaces, which radiate stored heat in cool weather and provide cooling in hot weather.

(continues)

(continued)

Four Horizons House has features that many other green homes may not need. Because of the area's high fire risk, the house has a steel roof and several steel structures. Wood collected for burning comes from the surrounding area to reduce the chances of fire spreading toward the house. Some of these fire-prevention features would be valuable in other hot, dry climates such as southern California.

Four Horizons demonstrates how a green building integrates with the environment around it while causing minimal damage to that environment. As an added attribute, Four Horizons offers a comfortable and serene home where a traditional house might be inappropriate. Some of the components of a house like Four Horizons require more effort, planning, and cost than conventionally built houses. Four Horizons has found a way to make up some of its building costs by expanding into individual units that now earn money as weekend eco-lodges for tourists.

Lindsay Johnson's Four Horizons House represented a new philosophy in building design: building in cooperation with nature rather than repressing nature. This concept laid the foundation for green building design. *(Ozetechture)*

A typical U.S. family of four uses about 350 gallons (1,325 l) of water per day at home. This does not account for additional water use outside the home at work or in school. Easy changes in behavior can reduce water waste, such as the following:

- showering instead of baths
- collecting the water that runs while waiting for hot water; use for watering plants, pet, etc.
- shutting off water in between each item when washing dishes
- saving laundry and dishes to make full wash loads
- planting drought-tolerant vegetation
- watering gardens only in the early morning or evening

Plumbing suppliers also offer a variety of products that lower water usage and water waste. The following table provides information on the most effective and commonly used water-saving devices for green buildings or conventional buildings.

Water utility companies in every U.S. community offer tips on how to save water inside and outside the house. Green buildings differ from traditional buildings because green builders pay extra attention to managing gray water and collecting rainwater. Gray water reuse has been gaining in popularity for several years, especially as a source of water for gardens. Some features of green buildings can conserve additional water by collecting rainwater and storing it in an aboveground or underground tank. Many green houses have cisterns, which are open tanks that simply collect any rain that falls into them. The water then runs to a storage tank. Architects often add features such as directed gutters along the roof's edge to carry rainwater to the cistern.

The practice of collecting rainwater for use is called rainwater harvesting. Rainwater is usually soft (contains few metals or salts) and clean. It requires minimal or no treatment before a household uses it for washing, dishwashing, or laundry. Rainwater to be saved for drinking should receive treatment by passing it through a filter installed between the storage tank and the taps. Treatment filters contain the following two components: a carbon filter that removes organic matter and a membrane filter that removes particles.

WATER CONSERVATION DEVICES	
DEVICE	HOW IT SAVES WATER
composting toilet	removes wastes to a composting area without reliance on water flushing
dishwasher	new countertop models can reduce normal dishwasher volume by almost half
dual-flush toilet	one volume for flushing solid waste and a smaller volume for flushing liquid waste
flash heater (on-demand water heater)	electric heating unit near the tap quickly heats small volumes of water, then turns off automatically when the flow stops
flow restrictors	constricted inner diameter of device allows less water to flow through, about 2.5 gallons (9.5 l) per minute
front-loading washer	reduces volume by one-third to one-half of top-loading machines, which use 8–14 gallons (30–53 l) of water per load
gray water reuse	rerouted wastewater from showers, sinks, and laundry rinse cycle goes to flush toilets or irrigation
low-flow showerhead and faucet	aerates water to lower the volume of the flow
low-flush toilet	reduces normal flush volume by half from 4–5 gallons (15–19 l) per flush to 1.6 gallons (6.1 l) per flush

MANAGING WASTE STREAMS

Water management within a green building means wastewater and other waste management. Waste management begins with the construction activities for a new green building and continues through to the daily routines of the inhabitants. Green builders have learned to use methods that reduce wood wastes and other excesses, but construction inevitably creates some waste materials. Good construction planning includes a list of sites where waste wood, concrete, stone, granite, fabric, and insulation may be sent for reuse.

Inside a green home, the owners manage their own waste streams, which are the total types and amounts of waste that the building produces in a period of time. The main waste streams consist of food wastes, paper and other recyclable materials, liquid wastes from toilets, washers, sinks, and showers, and solid human wastes. Most households reduce their waste loads by separating the recyclable materials to be picked up by a waste hauler. Green buildings include additional features to keep the other wastes from entering the larger community waste streams.

Many small and specialized companies offer products for reducing household waste. Composting toilets have been gaining acceptance as a safe way to reduce solid wastes and remove health hazards. Composting toilets work in either of two ways. First, a toilet can contain a receptacle that provides enzymes for breaking down the waste. Second, the toilet can treat the waste with enzymes and then direct the partially treated waste to an artificial wetland. Specialists can construct wetlands that contain a variety of plants and provide a slow but steady flow of water, both of which help natural microbes decompose the waste. Constructed wetlands therefore work exactly as natural wetlands work in decomposing organic matter.

Other techniques in waste reduction consist of an outdoor compost pile for nonmeat kitchen wastes, a gray water reuse system, and the collection of clothes for use in periodically reinforcing the house's insulation.

OFF THE ENERGY GRID

Individuals who are committed to living without overtaxing the Earth's natural resources have found inventive ways to exist off the energy grid. Even small communities have developed off-the-grid lifestyles through the cooperation of all the community's residents. Rock Port (population 1,300) in the northwestern corner of Missouri converted its energy use in 2008 to a completely off-the-grid system powered by four massive wind turbines. Skeptics in Rock Port doubted the town, even one as small as theirs, could go off the grid, but their land lies on the central plains where wind blows, and blows strong. A resident Eric Chamberlain, who led the conversion to off-the-grid living, admitted, "Did I ever think this would happen? Now, not in a million years . . . This is beyond my imagination." In high-wind seasons, Rock Port makes more electricity than it can use, so the town puts the excess on the municipal energy grid. In low winds, the town makes up the difference by drawing electricity from the grid. Rock Port's future may include the installation of energy storage systems so that the excess energy the turbines make can be saved for later.

Rock Port, Missouri, proves that off-the-grid living can take place in an entire municipality, although this town is small. Maybe the simplest way to move a larger and larger proportion of households and businesses from the energy grid rests in the hands of each individual. The environmental author Alex Steffen wrote in the 2006 book *Worldchanging: A User's Guide for the 21st Century*, "If houses with solar panels on their roofs and wind turbines in their backyards make you think of communes and hippies, your mental picture is out-of-date. Anyone with a bit of do-it-

This map shows the spots of highest energy use in the United States, correlating with high population centers. The map also suggests the challenges ahead for developing a national energy grid that is a smart grid, manages breakdowns in emergencies, and has the capacity to adjust to a growing population. *(NASA)*

yourself mindset and a little disposable income can benefit from installing a home-energy system. These setups can save you real money over the long term and provide most or all of your power in clean, homegrown ways." Put that way, there hardly seems a reason not to convert an existing building to some type of renewable energy source.

Advocates of wind power feel that establishing off-the-grid communities powered by wind may be cheaper and more feasible than installing solar energy systems. Rather than taking on the enormous job of converting large towns or cities, smaller communities of less than 10,000 people may be the best approach. The community wind advocate Mike Bowman told *E/The Environmental Magazine* in 2009, "We have a distribution system in this country where 80 percent of the geography is served by rural electrics. What we have today, 70 years later, is a system that's in place for delivering small amounts of power to thousands of places simultaneously." In other words, U.S. energy utilities already have a good distribution infrastructure for bringing renewable energy to thousands of small communities.

Bowman pointed out the advantages of community-scale wind power stations compared with large corporate wind farms. The following suggestions could apply just as well to utilities that supply solar power, geothermal power, or energy from biomass:

- Interconnected midsize installations can make better use of local geography than single large power plants.
- Most current transmission line grids do not have enough lines in the right places to carry electricity from solar or wind farms or geothermal sources.
- Plant managers could better control distribution and storage when the energy source fluctuates.
- Small systems can use existing power transmission lines.

Solar, wind, geothermal, and other renewable energies have established enough success stories in the United States and abroad to show that these off-the-grid methods are possible. They require only good planning, economic support—probably as tax credits—and a commitment from the community. The public has a bounty of resources at its disposal to make off-the-grid living practical. The best chance for success will likely be a strong economy that supports an innovative and growing green industry.

CONCLUSION

Green buildings provide the backbone of sustainable living in either large cities or tiny rural towns. Each building constructed today to use more solar energy and less coal-fired energy, more gray water and less city tap water, and more on-site recycling of wastes and less wastes sent to a treatment plant, helps build sustainability. Architects, designers, and environmental engineers all contribute to establishing these new green buildings.

Green building may have one of the brightest futures of all sustainability initiatives because new technologies in this area emerge frequently. Programs such as LEED help both homeowners and businesses by offering incentives to build green, and as an increasing number of towns require new buildings to be built green, the United States might move toward the once unthinkable status of being an off-the-grid society. Although the United States and other countries remain today very far from this goal, the desire and the technology becomes more focused every day on reaching some degree of sustainability in the near rather than the far future.

In order to take the big step to off-the-grid living, builders begin with simple steps in the construction of energy-efficient heating, cooling, and electrical systems. New technologies in insulation, windows, water recycling, and waste management support these systems. Green builders have excellent examples of homes that have transitioned from power-consuming to power-generating structures, and green building has become one of the fastest growing aspects of the construction industry.

The future of green building design will be led by the newest technologies in feedback mechanisms so that appliances, rooms, and entire structures can use energy at peak efficiency. This will be a major advance from the conventional energy distribution systems that most towns still rely upon, as described by the environmental writer Michael Prager in 2009, "The present grid has hardly changed in a century: massive amounts of power generated at behemoth plants are sent downstream via transmission wires. The system is stout and brawny . . . but it was never brainy, and don't even talk about its communication skills. (Have you ever considered that the only way the electric company knows your power is out is if you call and tell them?)" With this image in mind, it seems certain that scientists, engineers, and the public can improve on current energy production.

New technologies in building design, construction materials, energy-efficient methods of assembling new buildings, and a broadening array of renewable energy sources will be the future of green buildings. In many places, that encouraging future has already begun.

Energy from Solid Biomass

Liquid biofuels and solid biomass originate from matter that contains organic compounds. These substances are often referred to collectively as *bioenergy* sources. Most of the biofuel and biomass that have been envisioned as major future energy sources come from crops and crop residues left over after harvesting. Biofuels consist of mainly ethanol, an alcohol made from plant material (also called grain alcohol); the plant material from which ethanol is produced makes up biomass.

Ethanol and biodiesel are the two main biofuels in use today. A gallon (3.78 l) of ethanol contains about 67 percent of the energy supplied by a gallon of gasoline. Biodiesel comes from the processing of vegetable oils from various plants such as corn or soybeans or from vegetable fats. Biodiesel contains a different mixture of hydrocarbons than ethanol, so has a different quantity of energy: A gallon of biodiesel contains about 86 percent of the energy supplied by a gallon of gasoline.

Biofuels became the primary focus of the burgeoning alternative-fuel industry in the 1990s. As interest in new fuels and renewable energy sources bloomed, the worldwide investment in biofuels increased from $5 billion in 1995 to $38 billion on 2005, and it will top $100 billion by 2010. But the clamor for biofuels created unintended effects across the globe with an increase in corn prices—the main source of ethanol fuel—bigger than any increase farmers had seen since World War II. U.S. growers and farmers in other countries jumped at the chance to earn more by growing corn for the biofuel industry than stay with lower paying crops. From 2003 to 2008, the amount of U.S. corn planted has doubled.

High grain prices have meant increased prices of the items that use grains, such as beef, poultry, and breakfast cereal, among hundreds of other products. World food prices have begun to rise, and this rise has led to environmental harm. The scenario described by *Time* magazine in 2008 explained the effect of biofuels on the economy and in turn on the environment, as follows:

1. One-fifth of the U.S. corn crop diverts to ethanol refineries rather than food production.
2. The increased demand for corn raises world corn prices.
3. Extra land planted with corn makes the supply of other crops, such as soybeans, decline.
4. Soybean prices rise.
5. Farmers already growing soybeans in developing countries decide to increase their crop to take advantage of soy's rising value.
6. The farmers turn pastureland into cultivation, displacing ranchers.
7. Ranchers remove forests for more pastureland.

Forests that disappear in the name of biofuels equate to a loss of habitat for endangered species. The fallen trees additionally add large amounts of carbon dioxide (CO_2) to air that is already polluted as the ranchers burn whatever timber they cannot sell. Impoverished regions that cannot grow plentiful harvests of any kind fall victim to skyrocketing food prices. Growing corn for ethanol production furthermore consumes fuel for trucks and harvesters and for running ethanol refineries (called biorefineries). Cornell University's David Pemental said bluntly in 2007, "Biofuels are a total waste and misleading us from getting at what we really need to do: conservation. This [biofuels] is a threat, not a service." Global environmental organizations and economists like Nathaneal Greene of the National Resources Defense Council have now acknowledged, "We're all looking at the numbers in an entirely new way." A renewable energy source cannot have a future if it ultimately worsens poverty and devastates the environment. Studies on the worth of biofuels have continued. While environmentalists and some economists have identified the cautionary outcome to biofuel production, biofuel organizations and members of the federal government still support biofuel research.

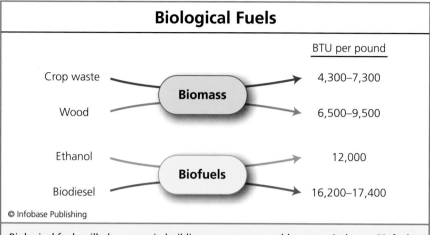

Biological Fuels

	BTU per pound
Crop waste → **Biomass** →	4,300–7,300
Wood →	6,500–9,500
Ethanol → **Biofuels** →	12,000
Biodiesel →	16,200–17,400

© Infobase Publishing

Biological fuels will play a part in building a strong renewable energy industry. Biofuels contain slightly more energy than most biomass fuels, but overall biomass may have lower costs regarding the energy needed to produce the feedstocks, the production process, and harm to the environment.

Biomass as an energy source has meanwhile developed in biofuel's shadow. The United States currently gets about 45 billion kilowatt-hours from biomass yearly, accounting for less than 2 percent of total electricity production. Energy from biomass may soon increase, however, because biomass does not interfere with current agricultural production, and it recycles the world's organic waste. Biomass as an energy source has already been shown to work, as this chapter discusses.

This chapter follows the growing importance of biomass as renewable energy. It defines biomass and compares it with other renewable energy sources. The chapter also describes the processes used in converting solid wastes to energy and finishes with a discussion on the future of biomass as a crucial energy source in sustainable communities. Finally, the discussion presented here offers ideas on how biomass may be optimized as a cheap, useful, and ecologically sound choice in energy production.

THE EARTH'S BIOMASS

Biologists think of biomass as the *dry weight* of all of the organic matter produced on Earth by plants and photosynthetic microbes. In environmental science, biomass is total plant materials but also animal wastes that can be burned as fuel.

Biomass is the energy-storage form for all living things in food chains. The chemical energy held in biomass serves each member of a food chain. For example, plant biomass in the form of carbohydrates provides energy to grazing animals; the biomass in these animals in the form of fats, proteins, and carbohydrates acts as the energy source for predators higher on the food chain. When animals produce waste or when they die, the biomass furnishes energy for microbes and for scavenger animals such as condors. Biomass therefore plays a central role in the Earth's nutrient recycling.

All of the Sun's energy stored on Earth in the compounds that make up plants and animals equals an ecosystem's *gross primary productivity* (GPP). When a plant or animal taps into this energy supply to live, grow, and reproduce, it must use some of the gross primary productivity for its own needs. Once those needs have been met, the energy left over is called *net primary productivity* (NPP), which is available for other organisms to use.

NPP = GPP − R, where R is the energy needed for an organism's systems

A portion of the energy in biomass disappears as heat whenever energy changes from one form to another. For example, a salmon in an Alaskan river consumes aquatic grasses for energy, but the fish cannot convert 100 percent of the plant energy into animal energy; some of the grass's energy dissipates as heat. Similarly, a grizzly bear feeding on the salmon can convert only a portion of the energy stored in the salmon's flesh. The rest of the energy also dissipates as heat. Such a stepwise scheme in food chain energy transfer is called an ecological pyramid. A large quantity of energy and organisms inhabit the bottom of the pyramid, but with each step to a higher level, the predators become less numerous and the energy available to them declines.

Activities on Earth convert biomass into energy in three different biological methods and one chemical method. In biology, microbes degrade biomass into simpler compounds with the release of heat and gases. The first microbial method is fermentation, which converts biomass to alcohols and other end products such as CO_2. The second microbial method entails anaerobic reactions, which are reactions that occur in the absence of oxygen. Anaerobic reactions produce mainly methane gas. The third biological method, respiration, is used by animals and some microbes. In respiration, an organism consumes oxygen as it converts sugars to energy and then releases CO_2 with other end products. The chemical method that occurs on Earth for releasing biomass's energy is combustion. A lightning

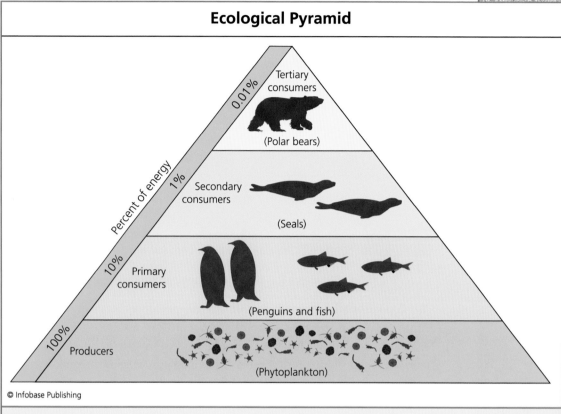

Ecological Pyramid

Percent of energy

0.01% — Tertiary consumers (Polar bears)

1% — Secondary consumers (Seals)

10% — Primary consumers (Penguins and fish)

100% — Producers (Phytoplankton)

© Infobase Publishing

Biology and chemistry must follow the laws of thermodynamics: (1) Energy cannot be created or destroyed and (2) some energy will be lost each time energy is converted from one type to another. An ecological pyramid illustrates energy use and energy loss. Each step up a food chain is associated with energy loss, usually as heat. A vehicle burning fuel works on the same principle: Most of the energy in fuel makes a vehicle run, but an amount of energy is always lost as heat.

strike may ignite a forest and cause the burning of dead leaves and branches as well as living trees. This burning converts the compounds making up biomass into different compounds with the release of heat energy. Making use of the energy that can be liberated from biomass through combustion is the basis of biomass energy production.

TYPES OF BIOMASS

Different types of biomass can be used for making energy in biomass power plants. When used in this manner for commercial or home energy production, the biomass materials are called feedstock. Feedstock originates from the following sources: agricultural crop waste (called bagasse), horticulture waste, wood and charcoal, pulp processing sludge, municipal solid waste (MSW), wastewater treatment solids, animal waste, and landfill waste. Sometimes used vegetable oils and animal fats also fit into the category of energy-producing biomass.

Biomass energy offers an advantage because it can be almost any solid material that when burned releases a usable form of energy. The main types of biomass used throughout the world differ in source so they contain various constituents, which make them more or less efficient as energy sources. Some of the variations in biomass are listed in the following table.

Civilization has used wood as its main biomass energy source for hundreds of centuries. In the United States in the 1800s, wood provided about 90 percent of energy use, but new energy sources replaced wood as new mechanized innovations came forward. Today, wood provides little more than 3 percent of the energy used in the United States. In developing countries, however, wood dominates all other energy sources, particularly in small rural communities. The Food and Agriculture Organization of the United Nations (FAO) estimates that more than 2 billion people worldwide fulfill their energy needs with wood.

TYPES OF SOLID BIOMASS WITH VARIABLE COMPOSITION	
SOLID BIOMASS	**POSSIBLE CONSTITUENTS**
agricultural waste	stalks, straw, cuttings, leaves, hulls, shells, vines, fruit and vegetable skins, seeds, animal manure
landfill waste and MSW	paper, cardboard, household garbage, restaurant waste, clothes and fabric, furniture
wood	pellets, chips and shavings, logging waste, branches, treetops, demolition waste, construction waste, cut timber, charcoal

Biomass energy production uses materials that an industry considers waste: crop residues, tree cuttings and trimmings, and wood scraps. Wood scraps make up an abundant biomass feedstock. In some timber applications, half of the tree goes into making a product and the other half is left as waste and can be used in energy production. *(National Wild Turkey Federation)*

Plant-derived biomass, such as wood, crop wastes, and paper, contains fibrous compounds that serve as the main storage form of the energy released in combustion. The three main fibers in plant biomass are lignin, cellulose, and hemicellulose, and materials high in these fibers are called *lignocellulosic biomass.* These three fibers vary quite a bit as evidenced by a woody log compared with a supple leaf from a grapevine. In general plant materials contain the following range of fibers: lignin 15–25 percent; cellulose 38–50 percent; and hemicellulose 23–32 percent.

Lignin provides strength to plant stalks and occurs at higher concentrations in woody materials. Burning biomass high in these fibers benefits

THE PHOSPHATE BOND

The metabolism inside cells of living organisms provides a good demonstration of how most work takes place in larger systems on the Earth. In order to move, communicate, and maintain their structure, cells and the larger organism that they compose must make energy. Just as important, cells must store energy until it is needed. Renewable energy systems work on the same principle: They make energy when fuel (sunlight, wind, steam, etc.) is available, and they should be able to store the energy in a form to be used later. Earth uses biomass as one of its main storage forms for energy; fossil fuels act as the other important storage form. Living organisms store their energy in a chemical structure called the phosphate bond.

The phosphate group (PO_4^-) consists of one phosphorus atom and four oxygen atoms, and it acts as one of six *functional groups* in biology. A functional group is a part of a molecule that participates in chemical reactions in the body. In human and all other animal cells, phosphate groups attached to other molecules transfer energy from one compound to another, but they also store energy when the cell rests. Adenosine triphosphate (ATP) serves as a critical compound that performs this energy storage and transfer with the aid of enzymes.

ATP holds three phosphate groups connected in a short chain. When an enzyme breaks the bond between the main portion of the ATP molecule and one of the phosphate groups, the reaction releases 7.3 kilocalories (kcal) of energy. This reaction equals about the same amount of energy as in a small bite of a candy bar. All of the more complex activities inside a living body begin with this first action of releasing energy from a phosphate bond.

Living cells convert fuel energy (food) into stored energy (fats, carbohydrates) and then store this energy for more immediate use in the phosphate bond of ATP and other compounds. The conversion of biomass adheres to the first law of thermodynamics: Energy can change from one form (chemical energy in biomass constituents) to another form (heat released in biomass combustion), but it cannot be created or destroyed. The human body, wildlife, microbes, plant life, and even fossil fuels all represent modes of transferring and storing the energy that the Earth receives from the Sun.

humans because they cannot digest these fibers as well as they digest carbohydrates such as starch and sugar. Since plants high in fiber do not serve well as food for humans, they make a good choice as an energy source for combustion. For this reason biomass is more attractive as an energy source than biofuels because, as discussed earlier, biofuels take land away from food production.

Biomass stores energy as chemical energy that is held mainly in the bonds between carbon and hydrogen. Combustion releases this energy in the form of heat by the following process:

biomass fuel + oxygen + heat to start the reaction → exhaust + heat

The first law of thermodynamics states that energy is neither created nor destroyed. In combustion of biomass, the energy created by the reaction equals the energy held by the constituents going into the reaction. (The sidebar "The Phosphate Bond" [page 163] explains how humans carry out this process.) The first law of thermodynamics therefore explains biomass energy production. Biomass power plants, sometimes called *waste-to-energy* (WTE) plants, convert the unusable form of energy held in biomass to a usable form. These usable forms may be heat, electricity, fuels for powering vehicles, or fuels for heating or powering buildings.

CONVERSION TO ENERGY AND FUELS

Biomass is a renewable energy because of its unlimited supply. Trees and plants regrow, animals give birth to young animals, and people continue to produce wastes. Biomass also offers several options as to how it can be used and the end products of its use. For example, the European Biomass Industry Association lists seven different processing methods for turning biomass into a usable end product, and the various end products can be biofuels, heat, electricity, chemicals, or another type of biofuel, such as the conversion of wood to charcoal.

In the United States, industry uses the greatest amount of biomass energy, almost 80 percent of total biomass energy production. About 20 percent of biomass energy goes to residential use and only 1 percent currently serves as feedstock for electric utility companies. Electric utilities would be wise to increase their dependence on biomass because biomass combustion is an uncomplicated process and similar to coal combustion that now produces most of the world's electricity. As mentioned, biomass also can be processed in a variety of ways so that a new power plant might choose a technology for biomass energy that works best in its circumstance. The following table describes predominant technologies for converting biomass into energy.

SOLID BIOMASS TECHNOLOGIES				
TECHNOLOGY	**PROCESS**	**DESCRIPTION**	**FEEDSTOCKS**	**PRODUCT**
aerobic digestion	biochemical	microbial digestion of sugars, followed by distillation	• crops • straw • wood • pulp	ethanol
anaerobic digestion	biochemical	microbial digestion of organic matter in a sealed oxygen-free tank	• manure • wastewater sludge • MSW	methane
biodiesel production	chemical	conversion to new hydrocarbons	• seeds • animal fat	biodiesel
direct combustion	thermochemical	burning	• agricultural waste • wood • MSW	• heat • steam • electricity
alcohol fermentation	biochemical	microbial digestion of organic matter	• agricultural waste • wood • paper	• ethanol • methanol
gasification	thermochemical	heating or anaerobic digestion	• agricultural waste • wood • MSW	• heating gas
pyrolysis	thermochemical	high-temperature treatment in absence of oxygen	• agricultural waste • wood • MSW	• synthetic oil • charcoal

Biomass energy production includes two technologies that will increase the overall efficiency of energy production. The first technology is called co-combustion or co-firing. In this process, biomass substitutes for a portion of coal being burned at a coal-fired power plant. Co-combustion might offer the following benefits: reduction of CO_2 emissions from coal; possible reductions in sulfur dioxides and nitrogen oxides, depending on the biomass composition; easy to modify existing coal plants; and abundant availability of biomass.

Cogeneration represents a second technology that biomass energy production may soon perfect. Cogeneration involves the simultaneous production of more than one fuel type, such as heat and electricity. According to the National Climate Change Committee of Singapore's National Environment Agency, newly built cogeneration plants afford an energy savings of 15 to 40 percent compared with conventional electric power plants in their power production operations. Most cogeneration plants in operation produce heat and electricity.

THE ENERGY VALUE OF GARBAGE

MSW that settles underneath new loads of waste in landfills has an energy value that should not be overlooked. Garbage—a familiar name for MSW—serves as an available form of solid biomass for making energy or fuels. Landfills contain very high numbers of microbes in the deepest layers where they decompose the organic materials. The decomposition that takes place in layers that hold little oxygen—anaerobic decomposition—leads to the formation of methane. Many landfills collect this methane and route it to energy utilities to be used the same as natural gas for heating and cooking.

Biomass recycling complements the natural paths of Earth's carbon recycling. Plants absorb from the air CO_2 exhaled by animals and convert the carbon to sugars that animals then use for food, and thus energy. When plant or animal life dies and decomposes, some of its carbon goes to CO_2 but some becomes trapped in sediments that sink into the Earth's mantle under tremendous pressure. After millions of years, the carbon turns into solid coal or it becomes liquefied due to the intense pressure and forms crude oil. Biomass is therefore part of an ancient process that has defined life and energy storage on Earth. The sidebar "Case Study: The Chicago Climate Exchange" on page 169 discusses how carbon cycling has in the 21st century been turned into a business.

The average American produces at least 4.5 pounds (2.0 kg) of biomass daily—about 1,600 pounds (726 kg) per year—in the form of simple garbage. The biomass naturally accumulates very rapidly, so burning it for energy seems to be an excellent option for both energy production and waste control. Today, the United States burns 14 percent of its solid waste in almost 100 WTE plants. About 1 ton (0.9 metric ton) of this garbage gives the same heat energy as 500 pounds (227 kg) of coal.

Many landfills install pipes that reach into the waste pile and collect the methane that anaerobic microbes produce. This methane has also been called biogas or landfill gas. The collection of methane produces a second energy-valuable use for solid biomass. The United States contains about 400 landfills that convert methane to energy for use by local communities. Depending on the size of the landfill, these operations generate enough power to furnish electricity to several hundred to a few thousand homes each year.

Wastewater treatment plants follow similar steps to capture the methane produced by microbes in the plant's anaerobic digester. The digestion step reduces the volume of excess sludge at a treatment plant and also produces energy-valuable biogas. Rod Bryden, executive at Plasco WTE facility in Ottawa, Canada, said at the opening of the plant in 2007, "The share of the waste that can be converted to power [by incineration] is not more than 18 to 22 percent. In ours we get about 44 percent to about 50 percent, a little more than twice as much power." WTE technology from landfills or wastewater treatment plants does not possess the glamour of new technologies in thin solar films or nanotechnology, but it contributes an important part of renewable resource energy use.

The following table summarizes the main advantages and disadvantages of today's biomass WTE technology.

A BIOMASS ECONOMY

The burning of biomass for energy production helps remove excess wastes from the following industries: agriculture, horticulture, forestry, and construction. It also helps destroy wastewater treatment plant solids and landfill contents, two materials that would otherwise have little value. By these activities, biomass energy production plays a role in the world's biomass economy.

Biomass economy refers to an accounting method for keeping track of the Earth's carbon compounds. This involves estimating where carbon

ENERGY PRODUCTION FROM BIOMASS	
ADVANTAGES	**DISADVANTAGES**
• removes accumulation of solid waste • large supply • makes use of otherwise unused timber, pulp and paper, and agricultural wastes • moderate to low costs • reduced CO_2 emissions • spares crops that can be used for food	• possible environmental damage from cutting forests • some emissions depending on composition and burning method • burning emits smoke and particles into the air

compounds are increasing and where they are decreasing. Before the industrial revolution, the atmosphere contained about 280 *parts per million* (ppm) of CO_2. As industrialization grew, machinery burned coal, natural gas, and oil, and the emissions from combustion drifted into the air. By the 1950s, CO_2 levels had reached 315 ppm; in March 2009 the atmosphere held 388.79 ppm. The CO_2 level increases about 2 ppm per year. CO_2 increases indicate that other greenhouse gases are also on the rise. Because greenhouses gases hold warmth in the atmosphere, the Earth's atmosphere is warming. In the IPCC report *Climate Change 2007,* scientists estimated that by the end of the 21st century global temperature will have increased 7.2°F (4°C).

More than 2 trillion tons of ice in Greenland, Alaska, and Antarctica have melted since 2003. The amount of water from only Greenland's melting ice in the last five years could fill 11 Chesapeake Bays. The rise in sea levels we face in the future is alarming. Thus, biomass energy production must be managed so that it helps remove CO_2 from the atmosphere rather than adds to greenhouse levels. Burning biomass produces CO_2. But if new plants grow faster than biomass is burned, the plant life can remove more CO_2 from the atmosphere than biomass burning puts into it.

Predicting where CO_2 levels are headed and how the elevated levels will hurt the environment is not easy. The world pours 8.8 million tons (8 million metric tons) of carbon emissions into the atmosphere a

CASE STUDY: THE CHICAGO CLIMATE EXCHANGE

The Chicago Climate Exchange (CCX) in downtown Chicago serves North America as a business that helps investors in the unique practice of buying and selling greenhouse gases. The CCX took shape in 2000 as an inventive means for reducing greenhouse gas emissions in the United States and worldwide by tying air pollution to the economy. The CCX does this through its main instrument, a procedure called *cap and trade*.

The cap-and-trade system accomplishes its goal based on two components: government-mandated caps on greenhouse gas emissions that industry must meet and the trading of permits that represent quantities of greenhouse gas emissions. The system begins with increasingly strict limits in industrial emissions. Each company holds a permit that identifies the amount of emissions it is legally allowed to put into the air. But as the legal limits become stricter over time, some companies or industries will have an easier time meeting the new requirements than others. As a result, certain companies produce emissions over their legal limit and other companies produce emissions under their prescribed limits. This discrepancy forms the basis of cap and trade.

In cap and trade, if CCX member Company A has exceeded its allowable emissions cap, it may purchase carbon credits from another member company, Company B, that has held its emissions under its own cap. For instance, Company B may have adopted renewable forms of energy such as solar panels or geothermal heating to help reduce emissions. As a consequence, Company B can make extra money by selling its carbon offsets to Company A. Company A then applies those offsets to its emissions total. This use of offsets theoretically brings Company A's total emissions under its legal allowable cap. The CCX brings together buyers of carbon offsets with sellers of carbon offsets. Offsets can be bought or sold for amounts of CO_2, methane, nitrous oxide, sulfur hexafluoride, perfluorocarbons (PFCs), and hydrofluorocarbons (HFCs).

By buying and selling carbon offsets, many people believe that the process will eventually reduce global warming by encouraging all industries to lower their emissions. The CCX founder Richard L. Sandor said

(continues)

(continued)

in 2003, one year after opening for business, "I'm more excited about the next 20 years in the environmental and social arena than I was about [other trading markets]. While it seems complicated to a lot of people, to me it's really simple." The CCX has since added more complexity to offset trading, but Sandor's original theory remains: The only way to get industry to cut emissions is to tie the cuts to profitability.

Europe now has a larger emissions trading exchange than the CCX, and other parts of the world have begun trading as well. No one yet knows if Sandor's idea will affect climate change, but new leaders often bring new commitment to the environment. An investment banker Seth Zalkin said in 2009, "With (President Barack) Obama taking office, it's going to start an accelerated process toward carbon regulation in the United States and a more vibrant market." The emissions trading market might soon join the New York and international stock exchanges as a bellwether for prosperity.

For success, the U.S. and international climate exchanges will require strict government controls similar to actions of the Securities and Exchange Commission (SEC) in governing the New York Stock Exchange (NYSE). Terry Barker and Igor Bashmakov warned in *Climate Change 2007*, an assessment report published by the Intergovernmental Panel on Climate Change (IPCC), that industries could learn to settle into a pattern of cap and trade that would demand little innovation from them toward reducing emissions. Trading on the CCX and other world carbon markets may eventually have an effect on global warming, but only if the volume of trades increases among a larger number of industries and includes effective government regulations on pricing and trading. Industries that buy credits must also work toward an overall decrease in their emissions or be held accountable. Without these actions, climate exchanges will be a success in theory only and not have a beneficial effect on the environment.

year. These emissions have altered ecosystems in some known ways and have undoubtedly caused hundreds or perhaps thousands of additional unknown alterations. Even with today's best technologies for reducing CO_2 in the atmosphere, humanity cannot save the environment from all the harm that comes from a growing population and expanding industry. The climate researcher Susan Solomon warned in 2009, "People have imagined that if we stopped emitting CO_2, the climate would go back to normal in 100 years, 200 years; that's not true." Environmental scientists must depend on new technologies that have not yet been invented to slow the rate of carbon buildup.

CONCLUSION

Biomass energy production serves as a simple technology for energy production when compared with advances in solar satellites, large tidal energy collectors, and other plans for the future of renewable energy. Biomass energy at its most basic is the collection and burning of organic wastes—not much different from what society has done for centuries. The new biomass energy industry will optimize this process by taking the best of conventional power methods and the best of alternative methods.

Biomass energy production helps reduce the world's tremendous buildup of wastes as well as lessen the need for burning fossil fuels. In other words, biomass energy accomplishes the objectives that society must meet to lower its ecological footprint. To make a difference in cleaning up the environment, biomass energy must avoid the mistakes made by earlier forms of energy, even biofuels. Biomass plants cannot rely on smoke-belching power plants that look like any coal-burning plant. Biomass planners must also lay out a scheme that allows agriculture to fulfill its main responsibility: food production. If the majority of farmers grow biomass crops instead of food crops, biomass will develop the same problems that biofuels have experienced.

The future of biomass requires that the biomass energy industry corrects the few drawbacks of this renewable energy form. First, biomass burning has the potential of producing air pollution. Biomass power plants will be expected to install scrubbers and other devices that remove gases and particles from emissions. Second, governments in several countries must assure that people do not begin cutting down forests for their biomass value. Destroying forestland removes critical habitat for endangered species, and the killing of trees releases huge amounts of additional CO_2 to the high levels already in the atmosphere. Finally, biomass energy production will conserve energy sources and natural resources in the most effective manner if it complements other forms of renewable energy.

FUTURE NEEDS

The subject of renewable energy contains many questions yet to be resolved. But innovations in renewable energy have emerged at an encouraging rate and continue to offer new approaches in energy use. The new ideas coming out of universities and small laboratories range from sophisticated programs built on elegant equipment to projects of rather modest technology but important all the same. For instance, under the umbrella of renewable energy, a student can choose among these technologies to pursue: space satellites that capture solar energy and beam it to Earth; unique power cells that use the energy systems of microbes; or artificial wetlands that occupy a small area in a backyard and decompose wastes as nature decomposes them. There hardly seems to be a discipline in science that offers the diversity of technologies found in renewable energy.

Ideas do not turn into reality without considerable sweat, and many of the goals in renewable energy have been set very high. There is pressure to implement the most promising technologies as soon as possible. A growing number of scientists have estimated how many years before fossil fuels will no longer be easily extractable, and others have calculated how many years before global emissions will cause permanent harm to the environment. The fact that we have reached a point where these predictions can be entered on a calendar indicates the dire condition the world is in.

Fortunately, few problems in the environment have been attacked with the fervor that science now has taken to renewable energy. The opportunities in renewable technology must be sorted and prioritized to be sure that the most feasible ideas are tried first, but also so that no seemingly farfetched idea becomes lost. Less than a few decades ago, few people would have imagined that surgeons could operate on patients using a laser, that a student could sit in a park and surf the Internet, or that a computer screen would show Parisians going about their day a half a world away. These ideas appeared only in science fiction; today they are real. Scientists and nonscientists only need to retain their faith in innovation to believe that inventions yet to be built can reverse the course for the Earth.

These thoughts of the future may be uplifting, but the environment needs help today. Governments and industries have opportunities to make some changes to energy production that will affect consumers within the next few years. A new administration in Washington, D.C., has already stated it wants progress to happen, and quickly, in many of the following areas: a modernized U.S. power grid and feedback capabilities; strict requirements for power companies to generate a portion of their product from renewable sources; monetary incentives to companies and households for saving energy; a plan for nuclear energy; continued commitment to lowering global greenhouse gas emissions, possibly by the cap-and-trade method; and substantive research that will replace old technologies with new to keep pace with the Earth's needs.

As in any long-term, technical, and expensive undertaking, no single segment of the population will get the work done by itself. Government will lead by setting and enforcing pollution standards; regulatory agencies will lead by holding polluters accountable for their wastes; industry will lead with new technologies; universities will lead by perfecting the most intensive approaches to energy use and production; and the public will lead as it has always led, by demanding change when change is needed.

The only thing renewable energy technology asks from everyone is a willingness to succeed. For the first time in history, a significant portion of the world's population now make decisions based on conservation and renewal. At least, for the first time, environmentalists have found people willing to listen who are not content with the old ways of dispensing electricity, heating a house, or powering a car. With a willingness to succeed, the renewable energy industry finally has a good chance of creating a sustainable future.

Appendix A

Top Oil-Consuming Countries in 2009	
Top Oil-Consuming Countries	Million Barrels per Day (MBD)
1. United States	20,698,000
2. China	7,855,000
3. Japan	5,041,000
4. India	2,748,000
5. Russia	2,699,000
6. Germany	2,393,000
7. South Korea	2,371,000
8. Canada	2,303,000
9. Brazil	2,192,000
10. Saudi Arabia	2,154,000
11. Mexico	2,024,000
12. France	1,919,000
13. Italy	1,745,000

Top Oil-Consuming Countries	Million Barrels per Day (MBD)
14. United Kingdom	1,696,000
15. Iran	1,621,000
16. Spain	1,615,000
17. Indonesia	1,157,000
18. Taiwan	1,123,000
19. Netherlands	1,044,000
20. Australia	935,000
Source: British Petroleum Oil Company	

Appendix B

PREDICTED WORLDWIDE ENERGY CONSUMPTION (QUADRILLION BTU)					
REGION	2010	2015	2020	2025	2030
North America	131.4	139.9	148.4	157.0	166.2
Asia	126.2	149.4	172.8	197.1	223.6
Europe	84.4	87.2	88.7	91.3	94.5
Central and South America	28.2	32.5	36.5	41.2	45.7
Middle East	25.0	28.2	31.2	34.3	37.7
Africa	17.7	20.5	22.3	24.3	26.8
Total World	509.7	563.4	613.0	665.4	721.6

Source: Timeforchange.org

Appendix C

INTERNATIONAL ORGANIZATIONS IN ENERGY		
ORGANIZATION	HEADQUARTERS	WEB SITE
International Atomic Energy Agency (IAEA)	Vienna, Austria	www.iaea.org
International Energy Agency (IEA)	Paris, France	www.iea.org
International Sustainable Energy Organisation (ISEO)	Geneva, Switzerland	www.uniseo.org
Organisation for Economic Co-operation and Development (OECD)	Paris, France	www.oecd.org
Renewable Energy Policy Project (REPP)	Washington, D.C.	www.repp.org
U.S. Department of Energy	Washington, D.C.	www.energy.gov
World Bank	Washington, D.C.	www.worldbank.org

Appendix D

MAIN RECYCLED MATERIALS	
RECYCLED MATERIAL	**EXAMPLE NEW USES**
aluminum	automotive; containers; keys
antifreeze	antifreeze
asphalt and asphalt mixtures	patching; road-base material; roofing
brick	crushed brick road-base material
coal ash	absorbents; concrete component
concrete and concrete mixtures	concrete mix; crushed concrete road-base material
cotton	pencils
computer disks	computer disks
ethanol	gasoline
fiberglass	acoustical ceiling panels; containers; custom molding
foam	insulation; paneling
glass	more than 50 products including beads; bottles; dinnerware; insulation; jewelry; paperweights; tiles; windows

RECYCLED MATERIAL	EXAMPLE NEW USES
hemp	clothing
metal	more than 20 products including cabinets; ceiling grids; clocks; containers; desktop organizers; food service items; furniture; jewelry; office supplies; signs; tableware; wind chimes
metal mixed with fiberglass, glass, paper, plastic, or rubber	flooring; furniture; thermometers
oil	diesel, fuel, gear, hydraulic, motor and tractor oils
organic materials	bags; bowls and plates; compost; food containers; mulch; plant mix; soil amendments
paint	latex; paint
paper	more than 100 products including absorbents; bags; books; boxes; calendars; computer paper; egg cartons; file folders; insulation; packaging; printing paper; stationery; tissue
paper mixed with glass, metal, plastic, textiles, or wood	binders; ceiling tiles; combs; marketing displays; pens; picture frames; rulers; snow scrapers
paperboard	paperboard
plastic	more than 300 products including aprons; automobile components; backpacks; bags; barriers; clothing; containers; cribs; food containers; furniture; insulation; mats; packaging; piping; playground equipment; posts; sheds; stadium seating; tiles; toys; truck bed liners; walkways; waste receptacles
plastic mixed with fiberglass, metal, paper, rubber, textiles, or wood	baseball caps; batteries; containers; lumber; mats; pens; playground surfacing; rulers; toner cartridges; walkways

(continues)

MAIN RECYCLED MATERIALS *(continued)*

RECYCLED MATERIAL	EXAMPLE NEW USES
rubber	more than 100 products including asphalt; bins; bumpers; flooring; liners; lumber; mats; railroad ties; speed bumps; surfacing; tires; turf
rubble	asphalt; cement; concrete
slag wool	insulation; paneling
steel	automotive parts; drums; tire gauges
textiles	absorbents; blankets; clothing; insulation; pencils and pens; pet beds; remnants
vinyl	bags; flooring; tags
wax	candles
wood	more than 50 products including animal bedding; benches; cabinets; custom molding; containers; decking; flooring; furniture; posts and stakes; railroad ties; wood chips
yard waste	lumber

Source: RecyclingMarkets.net

Appendix E

ENERGY VALUES USED IN ENVIRONMENTAL SCIENCE	
COMPARISON OF BTU CONTENT OF FUELS	
fuel	**Btu (British thermal units)**
1 barrel (42 gallons; 159 l) crude oil	5,800,000
1 gallon (3.78 l) gasoline	124,000
1 gallon diesel fuel	139,000
1 gallon heating oil	139,000
1 gallon propane	91,000
1 ton (0.9 metric ton) coal	20,169,000
1 cubicfoot (0.3 m^3 natural gas	1,027
1 KWh electricity	3,412
(1 barrel crude oil = 1,700 KWh electricity)	
FUEL QUANTITY TO MAKE 1 MILLION BTU OF ENERGY	
fuel	**quantity**
coal	90 pounds (40.8 kg)
gasoline	8 gallons (30.2 l)

ENERGY VALUES USED IN ENVIRONMENTAL SCIENCE *(continued)*

COMPARISON OF BTU CONTENT OF FUELS

natural gas	973 cubic feet (27.5 m^3)
wood	125 pounds (56.7 kg)

COMPARATIVE ENERGY VALUE OF A UNIT OF FUEL

fuel	Btu per pound (Btu/kg)
gasoline	20,192 (44,422)
biodiesel	16,211 (35,664)
coal	11,565 (25,443)
ethanol	11,471 (25,236)
wood	8627 (18,980)

APPROXIMATE ENERGY VALUES (1 BTU = 1,055 JOULES)

occurrence	energy value of the occurrence in joules
creation of the universe	10^{68}
supernova explosion	10^{44}
Earth in orbit	10^{33}
energy available from Earth's fossil fuels	10^{23}
yearly sunshine on the United States	10^{23}
U.S. energy consumption	10^{20}
earthquake, Richter 8.0	10^{18}
atomic bomb (Hiroshima, 1945)	10^{14}

APPROXIMATE ENERGY VALUES (1 BTU = 1,055 JOULES)	
occurrence	energy value of the occurrence in joules
putting space shuttle in orbit	10^{13}
1 U.S. resident energy use in a year	10^{12}
one-way Atlantic crossing of jet airplane	10^{12}
1 gallon (3.78 l) gasoline	10^{8}
explosion of 2.2 pounds (1 kg) TNT	10^{6}
1 candy bar	10^{6}
1 AA alkaline battery	10^{3}
human heartbeat	0.5
depressing a keyboard key	10^{-3}
1 photon of light	10^{-19}

Sources: University of Syracuse Physics Department; Oak Ridge National Laboratory

Glossary

ALGAE photosynthetic microbes common in freshwater and marine waters.

BATT building insulation made of synthetic fibrous material, such as fiberglass.

BENEFICIAL USE any use that is found for industrial wastes instead of burning or incinerating the wastes or putting the wastes into a landfill.

BIOENERGY any liquid or solid of biological origin that can be used as fuel for energy production.

BIOENGINEERING process of inserting genes from one organism into the DNA of a different organism for the purpose of giving the recipient organism new traits.

BIOFUEL a liquid or gas fuel made from plant material.

BIOGEOCHEMICAL CYCLE natural recycling of Earth's nutrients in various chemicals forms between living and nonliving things.

BIOMASS organic matter produced by plants and animals that can be used as a renewable energy source.

BIOME a terrestrial area defined by the things living there, especially vegetation.

BIOPRODUCTION the manufacturing of any item using biological rather than chemical materials and processes, usually referring to biofuels.

BRITISH THERMAL UNIT (Btu) unit of energy equal to the amount of heat required to change the temperature of 1 pound (0.45 kg) of water 1°F (0.55°C) at sea level.

CAP AND TRADE an economic means of controlling air pollution by paying companies that reduce their pollution with money from companies that exceed their legal pollution limits.

CARBON CYCLE natural recycling of Earth's carbon in various chemicals forms between living and nonliving things.

CARBON ECONOMICS a manner of keeping track of beneficial carbon compounds and environmentally harmful carbon compounds, usually for the purpose of monitoring carbon dioxide buildup in the atmosphere.

CARTEL an organization of political groups or businesses that work together to control the supply and prices of a product.

COAL-FIRED POWER PLANT an electricity-generating facility that burns coal to convert heat energy to electrical energy.

COMBUSTION process in which oxygen combines with other atoms to make a new compound and gives off energy as heat.

CONSERVATION well-planned and careful use of natural resources by humans.

DEVULCANIZATION process in rubber production or recycling that breaks chemical bonds so that the rubber polymers can be remolded into a new product.

DRY WEIGHT the weight of any substance after all the water has been removed.

ECOLOGICAL FOOTPRINT calculation of how much water and land a population needs to produce the resources it consumes and degrade the wastes it produces.

ECONOMY OF SCALE phenomenon in which products made by large production operations cost less per unit, in energy and in money, than when the same product is produced in a small operation.

ECOSYSTEM a community of species interacting with one another and with the nonliving things in a certain area.

ELECTROMAGNETIC SPECTRUM the entire range of radiation frequencies and wavelengths emitted by the Sun.

ERG a unit of energy equivalent to the force of 1 dyne exerted for a distance of one centimeter (cm). (1 dyne = force of accelerating 1 gram of matter at a rate of 1 cm/second^2.)

EXPONENTIAL GROWTH growth of some quantity at a constant rate so that the increases get bigger over time, for example, bacterial cells dividing at the following rate: 2, 4, 8, 16, 32, 64, 128, 256, and so forth.

FEEDBACK a feature of a system in which information flows from an end user back to the information source. A student answering a teacher's question demonstrates feedback.

FEEDSTOCK the raw materials entering a production process, such as corn for making ethanol fuel.

FERMENTATION   process carried out by microbes in which sugars are converted to alcohol and gas.

FISH LADDER   water-filled steps that enable fish that need to swim upstream to bypass barriers such as dams.

FOSSIL FUELS   products of the decomposition of plants and animals, exposed to intense heat and pressure in the Earth's crust over millions of years: coal, natural gas, and crude oil.

FUNCTIONAL GROUP   part of a molecule that participates in chemical reactions in the body.

GEOLOGICAL SURVEY   technique for finding fossil fuels in which a scientist studies the types, depths, and amounts of rock formations in a specific area.

GLOBAL WARMING   increase in the average temperature of Earth's atmosphere.

GRASSROOTS   describing a social movement that begins with individuals in local communities rather than from government.

GRAY WATER   excess water from sinks, showers, and laundry appliances that does not pose a health hazard and can be reused.

GREEN BUILDING   type of structure or construction process that involves activities to greatly reduce the consumption of nonrenewable resources and wastes.

GREENHOUSE GASES   gases in the Earth's lower atmosphere that hold in much of the heat rising from Earth and reflected from the Sun: carbon dioxide, methane, sulfur dioxides, nitrogen oxides, ozone, water vapor, and chlorofluorcarbons.

GROSS PRIMARY PRODUCTIVITY   all of the Sun's energy stored by the Earth's plants and animals.

HEAT EXCHANGER   an appliance that transfers heat from the indoors to outside in hot weather and transfers cool air from the inside to the outside in cold weather.

HUBBERT CURVE   a graph depicting the rate of oil production, which predicts the point in time in which the Earth will have reached peak oil production.

HYBRID VEHICLE   a vehicle that uses two different types of power source for the purpose of conserving fossil fuel.

HYDROCARBON   a long chainlike compound containing a carbon backbone with hydrogen molecules attached to each carbon.

HYDROELECTRIC POWER   electrical power generated from the kinetic energy in flowing water, usually dams.

HYDROKINETIC ENERGY   energy contained in the movement of water.

KINETIC ENERGY   energy contained in motion, such as wind or flowing water.

KYOTO PROTOCOL   an international treaty signed by many nations that agreed to lower emissions for the purpose of reducing global warming.

LIGNOCELLULOSIC BIOMASS   plant-derived biomass high in any of the fibers lignin, cellulose, or hemicellulose.

MEGAWATT (MW)   a unit of power equal to 1 million watts; a watt is a rate of energy production equal to 1 joule of energy per second. (1 joule = work done by the force of 1 newton through a distance of 1 m; 1 newton = force to accelerate 1 kg at a rate of 1 m/second^2.)

MEGAWATT-HOUR (MWh)   amount of work done measured as megawatts in one hour.

MICROBE   a single-celled microscopic organism such as bacteria and protozoa.

MINERAL WOOL   wool-like material in which the fibers are made of synthetic materials such as fiberglass or ceramic or stone.

NANOSCALE   of a size on the order of atoms or molecules.

NANOTECHNOLOGY   the science of synthesizing or manipulating devices that are the size of atoms or molecules.

NAPTHA SOLVENTS   liquids derived from the processing of petroleum.

NATURAL CAPITAL   Earth's natural resources that sustain life and economies.

NET PRIMARY PRODUCTIVITY   gross primary productivity less the energy needed for a plant or animal to live, grow, and reproduce.

NONRENEWABLE RESOURCE   a resource present on Earth in a limited amount that once it is used up the Earth cannot replenish for millions of years.

NUCLEAR ENERGY   energy generated from the fusion (combination) or fission (splitting) of atoms.

ORGANIC   pertaining to any matter containing carbon.

PARTS PER MILLION (ppm)   number of parts of a chemical found in a million parts of another material.

PHOTOVOLTAIC CELL   also solar cell; a device that converts radiant energy from the Sun to electricity.

PHYTOPLANKTON   tiny plant organisms that serve as food in freshwater and saltwater food chains, which absorb atmospheric carbon dioxide by photosynthesis.

POLYMER   a long chainlike compound usually containing a carbon backbone.

POWER GRID   also energy grid; the system for producing, storing, and distributing electricity (or natural gas).

PRIMARY RECYCLING   also closed-loop recycling; process of recycling a material or product to make more of the same material or product.

PROCESS WATER   water that has been used in manufacturing or other type of operation for washing or cooling equipment.

PROTOTYPE   an actual size nonoperating model of a device or vehicle for the purpose of demonstrating new features.

RADIOACTIVE   pertaining to an atom's action of emitting mass (alpha or beta particles or neutrons) or energy (gamma rays).

RECHARGING   the Earth's replenishment of a depleted resource, usually referring to aquifers and geothermal sources.

RENEWABLE ENERGY   energy from sources that do not deplete as they are used: wind, tides, hydropower, or solar.

RENEWABLE RESOURCE   a resource present on Earth in unlimited amounts or that can be replenished quickly compared to nonrenewable resources: water, geothermal resources, plants, forests, soil, animals, biomass, wind, or solar.

RESPIRATION   process in living cells of organisms in which glucose and oxygen are consumed and carbon dioxide is emitted.

RIPARIAN ECOSYSTEMS   ecosystems along streams, rivers, lakes, or coasts.

ROLLING BLACKOUTS   occurrences in which utility companies ration electricity to conserve a faltering power supply, leading to periods of low power supply to neighborhoods in sequence.

SCRUBBER   emission-reducing device that filters particles and some gases from industrial or vehicle emissions.

SECONDARY RECYCLING   also downcycling; process of recycling a material or product to make a new product.

SEDIMENT CYCLE   also rock cycle; the natural recycling of rock and soil by various physical and chemical changes from the Earth's surface to deep in the Earth's crust.

SOLAR CONCENTRATOR device containing materials that concentrates the energy produced by a solar cell, increasing overall energy output.

SOLAR PANEL an arrangement of many photovoltaic cells for collecting solar energy.

SPOT MARKETS short-term supplies of electricity available for purchase from utility companies from a large regional power grid.

SUSTAINABILITY ability of a system to survive for a finite period of time.

TECTONIC PLATE one of many various sized areas on the Earth's outer shell that move due to the fluid consistency in the Earth's mantle.

TERAWATT (TW) 10^{12} watts.

THERMAL MASS material that absorbs heat during warm periods and slowly releases heat as the surroundings cool.

THERMAL RESISTANCE material's ability to slow heat transfer.

TRANSESTERIFICATION a chemical step in the transformation of vegetable oil into biodiesel by replacing an alcohol functional group on the oil with a different alcohol.

TUNDRA a treeless plain of the arctic and subarctic regions characterized by permanently frozen subsoil and plant life consisting of mosses, lichens, herbs, and small shrubs.

TURBINE a machine that transfers one type of motion (wind or water) into another type of motion (rotating blades) in the process of turning kinetic energy into electrical energy.

WASTE-TO-ENERGY (WTE) any process that uses solid wastes or biomass as fuel to produce energy, usually as heat or electricity.

WATT-HOUR (Wh) amount of work done measured as watts in one hour.

WIND FARM operation containing many wind turbines for the purpose of generating large amounts of electricity.

ZERO WASTE the goal of sustainable activities in which all waste is eliminated or greatly reduced by making entirely recyclable products.

Further Resources

PRINT AND INTERNET

Al Husseini, Sadad. "Sadad al Husseini Sees Peak in 2015." Interview with Steve Andrews. ASPO-USA *Energy Bulletin* (9/14/05). Available online. URL: www.energybulletin.net/node/9498. Accessed January 11, 2009. Short interview with an oil expert on projections for future world oil production.

Alvarado, Mathew. "Tire Rejuvenation: Efficient ways to Devulcanize Industrial Rubber." Available online. URL: cosmos.ucdavis.edu/Archive/2007/FinalProjects/Cluster%208/Alvarado,Matt%20Tire_Rejuvenation4f.doc. Accessed January 6, 2009. Clear description of the technical aspects of rubber recycling and specific steps in the chemical process of reconstituting rubber.

Audubon. "Protect the Arctic Refuge." 2009. Available online. URL: www.protectthearctic.com. Accessed April 22, 2009. Opinions from the environmental group on the oil industry's effects on one of the last untouched ecosystems in the United States of America.

Bailey, Ronald H. *The Home Front: U.S.A.* Chicago: Time-Life Books, 1978. One of a historical series that reviews the U.S. economy in World War II.

Baker, David R. "Methane to Power Vehicles." *San Francisco Chronicle* (4/30/08). A news article on a company's plan to use methane to fuel its vehicles, with a brief description of the method.

———. "Electric Currents." *San Francisco Chronicle* (12/18/07). A clear description of wave power and its future with helpful diagrams of energy generation from wave action.

Barker, Terry, and Igor Bashmakov. "Mitigation from a Cross-sectoral Perspective." Chap. 11 in *Climate Change 2007: Mitigation of Climate Change*. Cambridge: Cambridge University Press, 2008. Part of a report recognized as a global resource for climate change policy, philosophy, and international agreements.

Bauman, Margaret. "Sen. Stevens Pushes for EAS, Alternative Energy Funding." *Alaska Journal of Commerce* (7/20/08). Available online. URL: www.

alaskajournal.com/stories/072008/hom_20080720008.shtml. Accessed January 12, 2009. An article that gives insight into Alaska's current progress in alternative energies in regard to the state's important oil economy.

Bourne, Joel K. "Green Dreams." In *National Geographic,* October 2007. An in-depth article on the advantages and disadvantages of ethanol biofuel.

Brinkman, Paul. "Obama Expected to Accelerate Efforts to Create U.S. Carbon Market." *South Florida Business Journal* (1/23/09). Available online. URL: southflorida.bizjournals.com/southflorida/stories/2009/01/26/story4.html. Accessed January 31, 2009. An article covering proposed changes to environmental programs as the Obama administration takes office.

BBC. "Bush Calls for Offshore Drilling." BBC News (6/18/08). Available online. URL: news.bbc.co.uk/2/hi/americas/7460767.stm. Accessed February 2, 2009. A news story explaining the political viewpoints on new U.S. oil drilling, especially in context of Republican versus Democrat viewpoints.

———. "1957: Sputnik Satellite Blasts into Space." BBC News: On This Day (10/4/08). Available online. URL: news.bbc.co.uk/onthisday/hi/dates/stories/october/4/newsid_2685000/2685115.stm. Accessed January 22, 2009. An article highlights a historical event in science that began satellite studies of the Earth and launched the U.S.-Russian space race.

Brown, Jeremy, and Pat Ford. "Science and Opportunity for Columbia/Snake Salmon." *Bellingham Herald* (12/18/08). Available online. URL: www.wild salmon.org/library/lib-detail.cfm?docID=794. Accessed January 14, 2009. An article on an environmental group's views on hydropower and the environment.

Bullis, Kevin. "Algae-Based Fuels Set to Bloom." Massachusetts Institute of Technology *Technology Review* (2/5/07). Available online. URL: www.livefuels.com. Accessed January 8, 2009. A discussion of the potential uses of algae farming and algae-based biofuels.

Bureau of Land Management. "Federal Agencies Move to Ease Development of Geothermal Energy and Increase Power Generation." News release (12/18/08). Available online. URL: www.blm.gov/wo/st/en/info/newsroom/2008/december/NR_12_18_2008.html. Accessed January 14, 2009. The BLM's announcement of plans to build a major geothermal energy project in the western United States.

Bush, George W. "State of the Union Address." (1/23/07). Available online. URL: frwebgate.access.gpo.gov/cgi-bin/getdoc.cgi?dbname=2008_record&docid=cr28ja08 -94. Accessed March 6, 2009. Presidential address to Congress and the nation on a variety of topics including the environment and oil drilling.

Cabanatuan, Michael. "Fueling a Revolution." *San Francisco Chronicle* (2/22/07). An article that explains small-scale biodiesel production and thoughts on the future of biorefining.

Capoor, Karan, and Philippe Ambrosi. *State and Trends of the Carbon Market 2008*. Washington, D.C.: World Bank Institute, 2008. Available online. URL: siteresources.worldbank.org/NEWS/Resources/State&Trendsformatted06 May10pm.pdf. Accessed December 9, 2008. The World Bank's annual report on the current size and trends in the global emissions exchange markets.

Cha, Ariana Eunjung. "China's Cars, Accelerating a Global Demand for Fuel." *Washington Post* (7/28/08). Available online. URL: www.washingtonpost. com/wp-dyn/content/article/2008/07/27/AR2008072701911_pf.html. Accessed January 8, 2009. Excellent update on China's role in current oil consumption.

Clary, Greg. "Recycling Converts Milk Jugs to Tax Savings." Lower Hudson Valley *Journal News* (1/13/06). Available online. URL: www.lohud.com/apps/pbcs.dll/ article?AID=/20060113/NEWS02/601130338/1017/NEWS01. Accessed January 6, 2009. This article gives an interesting perspective on a city's recycling economics.

Coase, Ronald H. "The Problem of Social Cost." *Journal of Law and Economics* 3 (1960): 1–44. Available online. URL: www.sfu.ca/~allen/CoaseJLE1960.pdf. Accessed March 6, 2009. A hallmark technical article that gave birth to the theory of carbon economics.

Emsley, John. "Making Viruses Make Nanowires to Make Anodes for Batteries." ScienceWatch.com (July/August, 2008). Available online. URL: science watch.com/ana/hot/che/08julaug-che. Accessed January 23, 2009. The latest technologies based on virus-produced batteries.

Fagan, Dan. "Who Will Obama Listen To? Alaska's Economy Hangs in the Balance." *Alaska Standard* (1/7/09). Available online. URL: www.thealaska standard.com/?q=node/248. Accessed January 12, 2009. Perspective from an Alaskan on climate change and the state's oil economy.

Federal Energy Regulatory Commission. "FERC Directs a Probe of Electric Bulk Power Markets." News release, July 26, 2000. Available online. URL: www. ferc.gov/industries/electric/indus-act/wec/chron/pr-07-26-00.pdf. Accessed December 6, 2008. A news release that represents the initiation of a corruption scandal in the U.S. energy industry.

First Solar. "First Solar Completes 10MW Thin Film Solar Power Plant for Sempra Generation." News release (12/22/08). Available online. URL: investor. firstsolar.com/phoenix.zhtml?c=201491&p=irol-newsArticle&ID=1 238556&highlight=. Accessed January 14, 2009. News release on a successful large-scale power plant using solar film technology.

Free Press News. "Science News: Some Climate Shift May Be Permanent" (1/29/09). Available online. URL: www.freep.com/article/20090129/NEWS 07/901290406/1009/Science+news++Some+climate+shift+may+be+

permanent. Accessed January 31, 2009. A science-based blog discussion on climate change and its consequences.

González, Ángel, and Hal Bernton. "Windfall Tax Lets Alaska Rake in Billions from Big Oil." *Seattle Times* (8/10/08). Available online. URL: seattletimes. nwsource.com/html/localnews/2008103325_alaskatax07.html. Accessed January 12, 2009. A news article on the effects the new Obama administration might have on Alaska's oil economy.

Grunwald, Michael. "The Clean Energy Scam." *Time* (4/7/08). An investigation into the mounting collection of hazards associated with growing biofuels.

Hillis, Scott. "Applied Sees Glass Solar Cell Demand Outgrowing Silicon." Reuters (3/19/07). Available online. URL: www.planetark.com/dailynewsstory.cfm/newsid/40936/story.htm. Accessed January 26, 2009. An update on the advancing technology of thin solar films.

Hotz, Robert Lee. "Make a Few Bucks, Help Fight Global Warming." *San Francisco Chronicle* (2/11/07). A news article that explains some of the debates regarding carbon trading.

Hubbert, M. King. "Energy from Fossil Fuels." *Science* 149, no. 2,823 (1949): 103–109. Available online. URL: www.hubbertpeak.com/Hubbert/science1949. Accessed January 13, 2009. A historical technical article predicting the peak of fossil fuel availability.

International Energy Agency. "New Energy Realities—WEO Calls for Global Energy Revolution Despite Economic Crisis." Press release (11/12/08). Available online. URL: www.iea.org/Textbase/press/pressdetail.asp?PRESS_REL_ID=275. Accessed December 10, 2008. An energy policy leader makes a powerful case for continuing sustainable energy programs despite poor economic times.

Johnston, David, and Kim Master. *Green Remodeling: Changing the World One Room at a Time.* Gabriola Island, British Columbia, Canada: New Society Publishers, 2004. Resource for materials and renewable energy systems in home building.

Johnston, Lindsay. "The Bushfire Architect." Interview with NineMSN.com (5/4/03). Available online. URL: sunday.ninemsn.com.au/sunday/feature_stories/transcript_1265.asp. Accessed March 5, 2009. Insight into the thinking of a green building architect and the originator of a world-famous off-the-grid house.

King, John. "S. F. Hopes to Set Example with New Green Tower." *San Francisco Chronicle* (4/13/07). An article that describes a power utility's plans to design and build an off-the-grid headquarters building.

Kunzig, Robert. "Pick Up a Mop." *Time* (7/14/08). An article on innovative technologies for reducing carbon dioxide levels, particularly technologies that affect ocean metabolism.

Lipp, Elizabeth. "Synthetic Biology Finds a Niche in Fuel Alternatives." *Genetic Engineering and Biotechnology News* (11/15/08). Clear explanation of the role biotechnology will play in synthesizing new alternative fuels.

Mabe, Matt. "Sun Is Part of the Plan for Greener Hempstead." *New York Times* (4/6/08). Available online. URL: www.nytimes.com/2008/04/06/nyregion/ nyregionspecial2/06solarli.html?_r=1&scp= 1&sq=greener+hempstead&st= nyt. Accessed January 14, 2009. Article on ways in which local communities convert to solar energy and the decisions that go into this process.

Maynard, Micheline. "Downturn Will Test Obama's Vision for an Energy-Efficient Auto Industry." *New York Times* (12/20/08). Available online. URL: www.nytimes.com/2008/12/21/business/21obama.html?scp=6&sq=auto mobile+industry&st=nyt. Accessed January 8, 2009. An update on the 2008 financial crisis in the American auto industry.

McCullough, Robert. Memo to McCullough Research clients, June 5, 2002. Available online. URL: www.mresearch.com/pdfs/19.pdf. Accessed December 6, 2008. A very interesting letter written at the time of the western energy crisis to shed light on the unlawful events of that period.

Morris, Frank. "Missouri Town Is Running on Vapor—and Thriving." National Public Radio *All Things Considered* (8/9/08). Available online. URL: www.npr. org/templates/story/story.php?storyId=93208355&ft=1&f=2. Accessed January 30, 2009. A short article on one town's ability to function without relying on the municipal power grid.

Mufson, Steven, and Philip Rucker. "Nobel Physicist Chosen to Be Energy Secretary." *Washington Post* (12/11/08). Available online. URL: www.washington post.com/wp-dyn/content/article/2008/12/10/AR2008121003681.html. Accessed January 12, 2009. A news article on the 2008 appointments by President-elect Barack Obama to federal energy positions.

National Security Space Office. "Space-based Solar Power as an Opportunity for Strategic Security" (10/10/07). Available online. URL: www.acq.osd.mil/ nsso/solar/SBSPInterimAssesment0.1.pdf. Detailed description of a next-generation solar power method using satellite solar cells.

North American International Auto Show. "Michelin Challenge Design Announces 2010 Competition Theme." News release, January 12, 2009. Available online. URL: naias.mediaroom.com/index.php?s=43&item=417. Accessed January 13, 2009. The latest news from the automotive industry on new car designs and concepts, presented from the world's premier auto show.

Olson, Drew. "Recycling an Old Argument about Recycling." *On Milwaukee* (4/9/08). Available online. URL: www.onmilwaukee.com/market/articles/ recyclingdebate.html. Accessed January 6, 2009. Interesting viewpoints on the opposition to recycling.

Pacific Gas and Electric. *Daylighting in Schools: An Investigation into the Relationship between Daylighting and Human Performance* (8/20/99). Available online. URL: www.pge.com/includes/docs/pdfs/shared/edusafety/training/pec/daylight/SchoolsCondensed8 20.pdf. Accessed January 26, 2009. Early evidence on the health benefits of sunlight in buildings.

Pollack, Andrew. "Behind the Wheel/Toyota *Prius*; It's Easier to Be Green." *New York Times* (11/19/00). Available online. URL: query.nytimes.com/gst/fullpage.html?res=9A02E7DA133BF93AA25752C1A9669C8B63&scp=1&sq=Prius&st=nyt. A historical article on the development of hybrid vehicles, based on the new Prius by Toyota.

Pomerantz, Dorothy. "Can This Man Save the World?" *Forbes* (8/11/03). Available online. URL: www.forbes.com/forbes/2003/0811/054.html. Accessed January 31, 2009. An article on the early days of carbon offset trading.

Popely, Rick. "Battery-Powered Car Race Is On." *Chicago Tribune* (6/18/08). Available online. URL: archives.chicagotribune.com/2008/jun/18/business/chi-wed-car-batteries_06-17jun18. Accessed January 9, 2009. News on the new generation of battery-powered cars from U.S. automakers.

Powell, Jane. "Green Envy." *San Francisco Chronicle* magazine (5/13/07). An editorial that highlights the pitfalls of green building trends.

Prager, Michael. "Gridlock. Real Energy Conservation Requires a Smarter Grid." *E/The Environmental Magazine* (January/February 2009). An opinion article on the current state of conventional power grids and their future.

Provey, Joe. "Building Wind Communities." *E/The Environmental Magazine* (January/February 2009). An article explaining ways to make wind power successful in U.S. communities.

Rabin, Emily. "Harnessing Daylight for Energy Savings." GreenBiz.com (4/18/06). Available online. URL: www.greenbiz.com/feature/2006/04/18/harnessing-daylight-energy-savings. Accessed January 26, 2009. Insight into the growing list of benefits offered by daylighting.

Rathje, William L. "The Garbage Project and 'The Archeology of Us.'" In *Encyclopaedia Britannica's Yearbook of Science and the Future—1997.* Edited by C. Cielgelski. New York: Encyclopaedia Britannica, 1996. Available online. URL: traumwerk.stanford.edu:3455/Symmetry/174. Accessed January 22, 2009. An article by a premier expert of the materials that make up municipal waste.

Reynolds, James. "Living in China's Coal Heartland." BBC News (1/22/07). Available online. URL: news.bbc.co.uk/2/hi/asia-pacific/6271773.stm. Accessed March 6, 2009. An interesting news report on the deteriorating health of people living in China's coal belt.

Richtel, Matt. "Start-up Fervor Shifts to Energy in Silicon Valley." *New York Times* (3/14/07). Available online. URL: www.nytimes.com/2007/03/14/technology/

14valley.html?pagewanted=1&_r=1. Accessed January 13, 2009. An article highlighting industry thoughts behind a new alternative energy industry.

Rosner, Hillary. "The Energy Challenge; Cooking Up More Uses for the Leftovers of Biofuel Production." *New York Times* (8/8/07). Available online. URL: query.nytimes.com/gst/fullpage.html?res=9B00E0DA1230F93BA3575BC0 A9619C8B63&sec=&spon=&pagewanted=1. Accessed January 12, 2009. An article offering insight into the disadvantages of biorefining.

Schwartz, Lou, and Ryan Hodum. "'Smart Energy' Management for China's Transmission Grid" (11/13/08). Available online. URL: www.renewable energyworld.com/rea/news/story?id=54061. Accessed March 6, 2009. A brief article on the status of China's bid to turn to smart energy distribution.

Science Daily. "Without Enzyme Catalyst, Slowest Known Biological Reaction Takes 1 Trillion Years" (5/6/03). Available online. URL: www.sciencedaily. com/releases/2003/05/030506073321.htm. Accessed January 9, 2009. A clear explanation of the role of enzymes in biology.

———. "Plastics Recycling Industry 'Starving for Materials.'" (10/16/07). Available online. URL: www.sciencedaily.com/releases/2007/10/071015111922. htm. Accessed January 6, 2009. A short article describing the pitfalls of plastic recycling.

Shah, Sonia. *Crude: The Story of Oil.* New York: Seven Stories Press, 2004. A well-written examination of the oil industry and the politics surrounding the manufacture and pricing of world oil supplies.

Solazyme. "Solazyme Showcases World's First Algal-Based Renewable Diesel at Governor's Global Climate Summit." News release (11/17/08). Available online. URL: www.solazyme.com/news081119.shtml. Accessed January 23, 2009. A news release explaining a new direction in fuel cell technology.

Steffen, Alex, ed. *Worldchanging: A User's Guide for the 21st Century.* New York, Harry N. Abrams, 2006. A classic reference for new technologies in sustainable living.

Stipp, David. "The Next Big Thing in Energy: Pond Scum?" *Fortune* (4/22/08). Available online. URL: money.cnn.com/2008/04/14/technology/perfect_ fuel.fortune/index.htm?postversion=2008042205. Accessed January 8, 2009. Background on the limited number of small companies that are growing algae for making biofuels.

Thompson, Elizabeth A. "MIT Opens New 'Window' on Solar Energy." *MIT News* (7/10/08). Available online. URL: web.mit.edu/newsoffice/2008/solarcells-0710.html. Accessed January 14, 2009. Explanation of how solar concentrators work for increasing energy conversion using solar power.

Tierney, John. "Recycling Is Garbage." *New York Times* magazine (6/30/96). Available online. URL: query.nytimes.com/gst/fullpage.html?res=990CE1DF133

9F933A05755C0A960958260. Accessed March 6, 2009. An oft-cited article that began the debate on whether recycling helps or hurts the environment.

Wald, Matthew. "New Ways to Store Solar Energy for Nighttime and Cloudy Days." *New York Times* (4/15/08). Available online. URL: www.nytimes. com/2008/04/15/science/earth/15sola.html?scp=1&sq=new+ways+to+store +solar+energy+for+nighttime+and+cloudy+days&st=nyt. Accessed January 14, 2009. An article providing information on new technologies for storing solar energy after collection until it is needed.

Walsh, Bryan. "Solar Power's New Style." *Time* (6/23/08). An update on the fast-growing technology in solar films.

Wittbecker, Greg. "Recycle to Save Energy—the Sooner the Better" (5/14/08). Available online. URL: www.stopglobalwarming.org/sgw_read.asp?id=23807 5142008. Accessed January 6, 2009. A short article that summarizes established energy savings from recycling.

Zero Waste Alliance. Available online. URL: www.zerowaste.org/case.htm. Accessed January 31, 2009. Resource on optimizing current recycling programs.

WEB SITES

AllPlasticBottles.org. Available online. URL: www.allplasticbottles.org. Accessed December 8, 2008. A simple coverage of plastics recycling and community programs.

American Solar Energy Society. Available online. URL: www.ases.org. Accessed January 30, 2009. Excellent resource for all aspects of solar power.

American Wind Energy Association. Available online. URL: www.awea.org. Accessed January 30, 2009. Excellent resource for all aspects of wind power.

Chicago Climate Exchange. Available online. URL: www.chicagoclimatex.com. Accessed December 9, 2008. The only North American market for carbon trading.

Electric Power Research Institute. Available online. URL: my.epri.com/portal/ server.pt. Accessed January 13, 2009. Good technical resource on energy technologies; access to the association's journal.

European Biomass Industry Association. Available online. URL: www.eubia. org/285.0.html. Accessed January 29, 2009. Resource for current technologies in biomass energy.

Federal Energy Regulatory Commission. Available online. URL: www.ferc.gov. Accessed January 13, 2009. Legal and government regulatory information on fossil fuel energy, electricity, and alternative energies.

Fuel Cells 2000. Available online. URL: www.fuelcells.org. Accessed January 9, 2009. Basic background on how fuel cells work and hydrogen chemistry.

Global Energy Network Institute. Available online. URL: www.geni.org/global energy/index2.shtml. Accessed December 10, 2008. An innovative organization with a plan for building a global energy grid.

GreenBuilding.com. Available online. URL: www.greenbuilding.com. Accessed March 5, 2009. An online resources for homeowners on sustainable materials, energy-saving systems, and LEED.

Greenpeace International. Available online. URL: www.greenpeace.org/international. Accessed January 14, 2009. An environmental group resource on the issues of alternative energy versus traditional fossil fuel and nuclear energies.

Intergovernmental Panel on Climate Change. Available online. URL: www.ipcc.ch. Accessed February 25, 2009. The primary resource on international policies and scientific reports on climate change.

National Recycling Coalition. Available online. URL: www.nrc-recycle.org. Accessed December 11, 2008. A good resource with data and tips, including an interactive recycling calculator.

Nuclear Energy Institute. Available online. URL: www.nei.org. Accessed January 14, 2009. A resource for the issues and advances in nuclear power.

Index

Note: Page numbers in *italic* refer to illustrations. The letter *t* indicates tables.

A

acid, from carbon conversion 119
acid rain 17
active solar heating 135
adenosine triphosphate (ATP) 163
air conditioners 138–139, 139*t*
Alaska 80–82, *82*, 84–85, *85*
algae 61–62, 123
alternative fuels 52, 56*t*, 61, 77, 78*t*, 89, 91, 119, *120. See also* battery power; biodiesel; biofuels; fuel cells; natural gas; synthetic fuels
alternative fuel vehicles 51–74
    battery power for 64
    biofuel for 57–62
    design of 59
    development of 55–56
    evolution of 53–57, 54*t*
    fuel cells for 66–69
    hybrid vehicles *55,* 64, 66, 72–74
    natural gas for 70–72
    next generation of 72–74
    popularity of 52
    solar-powered 53
    synthetic fuels for 62–64
    trucks 56–57
aluminum industry 34
aluminum recycling 34, 36–37, 49

Alyeska pipeline. *See* Trans-Alaska Pipeline System
American Institute of Architects (AIA) 127–128
anaerobic decomposition 166
anaerobic reactions 159
antifreeze 135–137
anti-pollution technology 57
appliances, smart 23
Ausra 103

B

bacteria, for fuel cells 123
Baku-Tbilisi-Ceyhan pipeline 83
balanced ventilation 139
battery power 64–66, 72–73, 123
behavior 130–134, 149
binary power plants 111
biodiesel 61, 87–90, *89,* 123, 156
biodiesel blends 87–88
bioengineering 64
bioethanol. *See* ethanol
biofuels 53–55, 57–62, *60,* 74, *90,* 123, 156–157, *158. See also* biodiesel; ethanol
biogas 72–73, 167
biological fuel cells 66–69, *67,* 121, 122–123

biological insulation 140
biomass 63, 68–69, 74, 86, 158–164,
    161t, 166–167, 171
biomass economy 167–171
biomass energy *158,* 158–171
    advantages of 168t, 171
    carbon dioxide from 20
    by combustion 164
    disadvantages of 168t, 171
    feedstock for 161, *162,* 166–167
    industrial use of 164
    methods for 159–160, 165t
    used by utilities 164
bio-oil 86
biorefining 75–92
    by-products of 88, 91
    definition of 75
    future of 91–92
    obstacles to 91
    pilot plants for 89–90
    steps in 86–89, *87*
biorefining industry 89–91
birds, and wind turbines 99
Brazil 58–59
British Petroleum 109
Btu (British thermal unit) 3
buildings 125. *See also* green building
    design

C

California energy crisis 6–7
Canada 80
cap-and-trade system 169
Cape Cod (Massachusetts) 97
carbon 16–17, 63, 118–119
carbon dioxide 17, 20, 119, 168. *See also*
    emissions; greenhouse gases
carbon economics 16–20, *19,* 169–170
carbon fuel cells 121
carbon offsets 17–20, *18,* 169–170
carbon sequestration *120*
carbon trading 17–20, *18,* 169–170
cardboard 140
caribou 84–86
cars. *See* vehicles
catalysts 67

catalyzed reactions 43
cellulose 162
cellulose insulation 140
cellulosic ethanol 61
central ventilation 139
CFC (chlorofluorocarbon) 140
chemical fuel cells 121–122. *See also*
    hydrogen fuel cells
Chernobyl accident 115
Chicago Climate Exchange 17–20,
    169–170
China 14, 25, 51, 97–98
Chrysler 59
cisterns 149
clean energy. *See also* direct carbon
    conversion; fuel cells; geothermal
    energy; nuclear energy; solar energy;
    wind power
    calculating 93
    definition of 93
    future of 124
    hydropower 99, 109–110, 111t
    innovations in 93–124
    need for, factors in 94–95
    tidal power 99–101, 99t, *100*
    wave power 99–101, 99t
climate change. *See* global warming
Climos 118
closed-loop recycling 35
coal 22, 63, *79*
coal mines, for gas storage 83
coal mining 5
coal-to-oil processing 80
Coase, Ronald 17–18
co-combustion (co-firing) 166
cogeneration 166
combustion 65, 159–160, 164
composting toilets 151
compressed natural gas (CNG) 72
conduction 140
conservation. *See* energy conservation;
    water conservation
consumers 124, 173
consumption gap 13–14
convection 140
conventional natural gas 70

cooking oil 61

cooling systems 138–139, 147

corn 60, 74, *90,* 156–157

crops, biofuel 57–61, 74, 156–157

crude oil 10–16, *11, 12t,* 75–76, 77. *See also* oil reserves

crumb rubber 48

curbside recycling 33, 37–38

D

dams. *See* hydropower

daylighting 142–144, 147

Daylighting Initiative 143

deforestation 157, 171

Department of Energy (DOE) 81

devulcanization 47, 48–49

diesel alternatives 61

diesel vehicles 57

direct carbon conversion 117–120, *120*

distillation 42

downcycling 35

Druzhba pipeline 82–83

dry steam 110

ductless air conditioners 138–139

E

Earth Day 34

ecological footprint *4,* 4–6, 10, 30

ecological pyramid 159, *160*

economics 3–4, 7, *8,* 45–48. *See also* carbon economics

economies 45–48, 84–85, 157, 167–171

electrical energy, from Sun 101–102. *See also* solar energy

electric-gasoline hybrids *55,* 72

electric power 3, 6–7. *See also* clean energy

electric supply 6–7

electric vehicles 53

electromagnetic radiation 15–16

electromagnetic spectrum 15, *15t*

emissions 22, 57, 97, 116, 118, 168–171

emissions trading. *See* carbon trading

energy brokers 6–7

energy, clean. *See* clean energy

energy conservation 1, 2, 31–32, 35–39, 43, 130–134, 144, *145t*

energy consumption 2–7, *3t, 5, 8,* 25, 27, 28–29, 30

energy conversion 159

energy grids 6, 22–24, *23,* 26–27, 151–153, *152, 154*

energy programs, global 26–27

energy sources *3t,* 5, 24–26, 93. *See also* nonrenewable energy sources; renewable energy sources

EnergyStar 138, *139t*

energy sustainability 29

energy systems 129–138, 147

Enron 6–7

entrepreneurs 39, *39t,* 40

environment 28, 94, 127, *143*

environmentalism 34

Environmental Protection Agency (EPA) 93

enzymes 64, 67–68, 122–123

ethanol 57–59, 61, 74, *90,* 156–157

EV-1 (GM) 64

evaporative coolers 138

exhaust-only ventilation 139

exploration, natural gas 70, 71

F

feedback, in energy grids 23

feedstocks 57–59, *60,* 63, 88, 89, 91, 161, *162*

fermentation 64, 159

filtered water 149

filtration 42

fire retardants 140–142

first law of thermodynamics 163

First Solar 109

Fischer-Tropsch process 63

fission reactions 69–70, 114

flash steam plants 111

flex-fuel vehicles 56–57, 73

food chain 16, 159

Food, Conservation, and Energy Act (2008 Farm Bill) 90

food production 157, 162–164, 171

Ford 59

fossil fuels 17, 29. *See also* coal; crude oil; natural gas
Four Horizons House 147–148, *148*
fuel cells 66–69, *67, 68,* 72–73, 74, 117, 120–123, 122*t*
functional groups 163
fusion reactions 16, 69–70

G

gamma rays 16
garbage 34, 166
Garbage Project 34
gasoline-battery hybrids 64
General Electric (GE) 97–98
General Motors (GM) 59, 64, 66
geothermal energy 110–113, 111*t, 112,* 114*t*
geothermal exchanger 111–112
geothermal heat pumps 111, 135
glass recycling 38–39, 41*t*
Global Energy Network Institute 26–27
global warming 17, 20, *26,* 55, 65, 81, 84, 168–171
glycerin 88, 91
GM (General Motors) 59, 64, 66
GM EV-1 64
GM Volt 66
grassroots environmentalism 34
Grassroots Recycling Network 35
gray water 149
Greasecar 61
Green Building Council 131
green building design 125–155, *128*
    certification for 131–134
    cooling systems in 138–139
    daylighting in 142–144
    decisions in 126–127
    definition of 126
    energy systems in 129–138
    environment and 127, *143*
    focus areas of 126
    of Four Horizons House 147–148, *148*
    future of 154–155
    guidelines for 128
    heating systems in 129–138, 136*t*
    insulation in 140–142, 141*t*
    objectives of 126
    solar energy in 134–137
    for sustainability 154
    technologies developed for 154
    today 128–129, 130*t*
    trendiness of 128–129
    ventilation systems in 138–139
    waste management in 150–151
    water conservation in 146–149
    windows in 144–146
green building movement 125–129
green chemistry 64
GreenFuel Technologies 61
greenhouse gases 17, 52, 93, 116, 118, 168–171
Greenpeace 114–115
gross primary productivity (GPP) 159
Grove, William 121

H

Harrop, Peter 108
Harvard University 135–137
HCFC (hydrochlorofluorocarbon) 140
HDPE (high-density polyethylene) 42–43
heat 140, 144, 159
heating systems 129–138, 136*t,* 147
heat pumps, geothermal 111, 135
hemicellulose 162
hot dry rock 112
hot rock technology 112
hot water energy 110
house(s) 129, 132, 133*t,* 135, *137*
Hubbert, M. King 75–76
Hubbert Curve 95, *95*
hybrid vehicles 55, 64, 66, 72–74
hydrocarbons 10, 63, 64, 65
hydrogen fuel cells 66–69, *68,* 74, 121
hydrokinetic energy 99
hydropower 6, 22, 99, 109–110, 111*t*

I

IEA (International Energy Association) 27
India 51

industrialization 4, 25, 51, 168
industrial materials 39–41, 40*t*
industrial recycling 40–41, 40*t*
industry 71, 173
insulation 135, 140–142, 141*t*
internal combustion engine 65, 94
iron-seeding 118

J

Japan 58
jatropha 61
Johnston, Lindsay 147–148

K

kinetic energy 22
Kyoto Protocol 20

L

landfills 31, 72, 166, 167
land, for geothermal plants 113
Leadership in Energy and
    Environmental Design (LEED) *131,*
    131–134, 133*t*
lifestyles 3–4, 151–153, 154
lighting design 142–144
lignin 162–164
lignocellulosic biomass 162–164
limpet 100–101
liquefied natural gas (LNG) 71–72
lithium-ion batteries 64–66
low-carbon energy 27
Low-E windows 145

M

magma 112
Massachusetts Institute of Technology
    (MIT) 108
materials chemistry 39, 41–44
materials recovery centers 35–36
materials science 39
metal 44*t*, 45, 67
metal recycling 41*t*, 45–48. *See also*
    aluminum recycling
methane 17, 72, 166, 167
Michelin Challenge Design 73–74
microbes 64, 122–123

mineral 44
mineral recycling 44–45
mining 5, 44–45
motor oil, synthetic 63
municipal recycling programs 33–34
municipal solid waste (MSW) 166

N

National Security Space Office (NSSO)
    105–106
natural capital 1. *See also*
    nonrenewable resources; renewable
    resources
natural gas 70–73, 74, 83
natural gas liquids 71
natural gas pipelines 22, 83
natural light 142–144
natural ventilation 139
naval ships 69–70
net primary productivity (NPP) 159
newspaper 140
nonrenewable energy sources 68, 113
nonrenewable resources 1, 2, 8, 9*t*
nuclear chain reactions 69
nuclear energy 69–70, 113–117
nuclear power 115–116, 115*t*
nuclear reactors 114, 116–117

O

ocean, carbon absorption 118–119
off-the-grid living 151–153, 154
oil. *See* crude oil
oil consumption 13–14, 52, 75, 76–77
oil deficit 78
oil fields, depleted 76, 83
oil industry 76, 77–80, 84–85
oil pipelines 82–86
oil refining 10–11, 77–78, 79–80
oil reserves 12–13, 75, 77, 78
OPEC (Organization of the Petroleum
    Exporting Countries) 94, 95*t*

P

Pacific Gas and Electric 143
paper recycling 33, 34–35, 41*t*
passive solar heating 134–135

peak energy usage 23, *24*
peak oil production 95
peak oil supply 75–76
pelamis 100–101
petrochemicals 10–11
petrochemistry 13
phase separation 42–43
phosphate bond 163
photosynthesis 16
photovoltaic cells 101–102, *102,* 105
phytoplankton 118
pilot biorefineries 89–90
pipelines 22, 80–86
plant biomass 162–164
plastics 38–39, 41, 41*t,* 42–43, 49
plug-in hybrids 66
pollution, in China 25
polyethylene 42–43
population 8–10, 30
poverty, and sustainability 25
Powell, Jane 128–129
power buoy 100–101
power plants 22, 23, 110–117, *112, 116,*
    161, 164, 167
primary recycling 35
Prius 53, 58–59, 64–66
prototype vehicles 55, *73*
Public Utilities Commission 137–
    138
pure biodiesel (B100) 87
pyrolysis 86

R

radiation 140
radioactive waste 114, 117
rainwater harvesting 149
Rathje, William 34
recyclable materials 32, 32*t,* 37
recycled materials 37, 39–41, 43
recycling 31–50. *See also* aluminum
    recycling; glass recycling; industrial
    recycling; metal recycling; mineral
    recycling; paper recycling; plastics;
    rubber recycling
        cost of 37
        criticism of 31, 37–38

efficiency in 37–38, *38*
energy conserved by 31–32, 35–39
history of 33–35
innovations in 39*t,* 40
process of 35–36, *36,* 41, 41*t*
in sustainability 31
in waste management 49
during World War II 46–47
recycling chemistry 41–44
recycling industry 40
reduce, reuse, recycle 127
renewable energy sources 20–22, 21*t,*
    30, 61–62, 81, 124, 164, 172–173
renewable resources 2, 8–10, 9*t*
resin 42–43
resin codes 42–43
resource management 2
respiration 159
Ring of Fire 113
Rock Port (Missouri) 151–152
rolling blackouts 6, 7
rubber recycling 46, 48–49
R-value 140

S

salmon 110
salt caverns 83
school lighting 143
scrap drives *46,* 46–47
scrap tires 48, *48*
scrubbers 119
sea levels 168
seawater, chemistry of 118–119
secondary recycling 35
semiconductors, virus-based 123
Shah, Sonia 12–13
smart appliances 23
smart energy grids 22–24, *23, 24, 152*
Solana *104,* 104–105
solar cell glass 146
solar cells. *See* photovoltaic cells
solar collectors 135
solar concentrators 105, 108
solar energy 101–108
        advantages of 103*f*
        capturing 101–102, *102*

demand for 102
disadvantages of 103*f*
in Four Horizons House design 147
government policies on 103–104
for heating systems 134–137
obstacles to 102–103
price of systems 24–25
satellite collection of 105–106
storage of 102–103
technologies for 107*t*
solar films 108–109
solar films (window) 146
solar heat gain coefficient 145
solar panel arrays *104,* 104–105
solar panels 105, *106*
solar-powered vehicles *53*
solar satellites 105–106
solar thermal 103
Solazyme 123
stacked fuel cells 69
sugarcane 60, *60*
Sun, energy from 14–16, 101–102, 106. *See also* solar energy
sustainability 1, 2, 25, 29, 30, 31, 59, 154
synthetic biology 63–64
synthetic fuels (synfuels) 62–64, 65
synthetic motor oil 63

T

tar sand processing 80
thermodynamics, first law of 163
thermoplastics 42
thermoset plastics 42
Three Rs 127
tidal power 99–101, 99*t, 100*
tire recycling 48
tires, scrap 48, *48*
Toyota Prius 53, 58–59, 64–66
Trans-Alaska Pipeline System (TAPS) 80–82, *82,* 84–86, *85*
transesterification 88
transportation 52. *See also* vehicles
trout 110
trucks 56–57

U

unconventional natural gas 70
United States. *See also* Alaska
alternative fuels in 55–59, 77
energy consumption in 2–4, 3*t, 152*
energy crisis in 6–7
energy innovations in 173
energy policies of 81
geothermal energy projects in 113
HDPE recycling in 43
hydropower in 110
natural gas pipelines of 22, 83
oil in 11, 12*t,* 13–14, 76–77, 78, 95
radioactive waste storage in 117
recycling in 32*t,* 45–48
smart energy grids and 23–24
vehicle usage in 51
wind energy used in 98
United States Department of Energy. *See* Department of Energy (DOE)
United States Environmental Protection Agency. *See* Environmental Protection Agency (EPA)
United States Green Building Council 131
United States House of Representatives Subcommittee on Energy and Environment 28–29
uranium 113–114
U-value 145

V

vegetable oils 61, *87*
vehicles 51–52, 59, 64, 65, 94. *See also* alternative fuel vehicles
ventilation systems 138–139
viruses, for fuel cells 123
Volt (GM) 66
vulcanization 47

W

warm-rock reservoirs 112
waste management 34, 49, 150–151

Waste Management, Inc. 72
waste-to-energy (WTE) plants 164
wastewater treatment plants 167
water 109–113, 149
water conservation 146–149, 150*t*
water management 150–151
wave power 99–101, 99*t*
wells 112–113
wetlands 151
wet steam 110
windows 135, 144–146, 145*t*

wind power 97–99, *98, 99t,* 151–153
wood, as energy source 161
wood furnaces 135
World War II 46–47
Wright, Allen 119

Y
Yucca Mountain 117

Z
zero waste 35, 50